AF734229

"Success is not final, failure is not fatal: it is the courage to continue that counts."

—Winston Churchill

This book is dedicated to all those who have inspired and supported me throughout this journey.

Foreword

The field of intelligent and efficient video moment localization has witnessed significant advancements in recent years, driven by the increasing demands for precise and scalable video analysis. This monograph provides a thorough and insightful exploration of the latest developments, seamlessly integrating foundational concepts with state-of-the-art innovations. By addressing both theoretical and practical aspects, this work serves as a valuable resource for researchers, practitioners, and students in multimedia analysis, video processing, and artificial intelligence.

As we continue to push the boundaries of this rapidly evolving domain, this book stands as a timely and essential reference. Its comprehensive coverage and clear presentation ensure that it caters to both experienced experts and newcomers seeking to understand the complexities of video moment localization. We are confident that the strategies, methodologies, and insights presented in this book will inspire further research and innovation, helping to shape the future of video understanding and its applications in the real world.

Jinan, China — Meng Liu
Jinan, China — Yupeng Hu
Shenzhen, China — Weili Guan
Shenzhen, China — Liqiang Nie
January 2025

Preface

The exponential growth of digital content and the rapid expansion of video-sharing platforms have led to an unprecedented accumulation of videos, spanning domains such as surveillance, education, entertainment, and social media. While this rich video archive has revolutionized how we capture, share, and consume experiences, it also poses significant challenges in efficiently and accurately retrieving specific content. This challenge becomes particularly evident when users seek to pinpoint precise moments within untrimmed and unsegmented videos using natural language queries. Video moment localization, a rapidly emerging research field, aims to address this challenge by enabling intuitive and flexible retrieval of video content.

Conventional video analysis approaches, including action recognition and temporal action localization, are often limited to predefined action categories or require labor-intensive, moment-level annotations. These constraints hinder their ability to address real-world needs, where flexible, scalable solutions are required to process large-scale, unstructured video datasets and handle complex, free-form queries. The introduction of natural language-based video moment localization represents a pivotal advancement, allowing users to search for specific moments using natural, human-like queries while bridging the gap between video understanding and semantic reasoning.

However, despite notable progress, several critical challenges persist. These challenges include understanding diverse spatial-temporal characteristics of video content, achieving precise cross-modal semantic alignment between textual queries and video data, managing computational efficiency, and scaling solutions to handle vast video corpora. Additionally, the reliance on supervised learning and annotated datasets further limits the adaptability of existing methods to real-world, dynamic scenarios.

In response to these challenges, this book provides a comprehensive exploration of the latest advancements in video moment localization, addressing the key obstacles in accuracy, efficiency, and scalability. Specifically, we present innovative approaches for contextual understanding and cross-modal semantic alignment, including methods such as semantic-enhanced and semantic alignment-based video moment localization, which utilize advanced attention mechanisms and dynamic query decomposition to improve understanding between visual and textual modalities. Furthermore, we address computational efficiency and scalability through techniques like semantic pruning and efficient hashing-based localization, enabling large-scale retrieval while reducing computational costs. To foster deeper integration of modalities, we introduce frameworks like semantic collaborative localization, which facilitate effective information fusion between visual and textual representations. Recognizing the high cost of labeled data, we also explore weakly supervised learning approaches that rely on video-level annotations, significantly reducing training overhead while maintaining strong performance.

Beyond detailing current methodologies, this book examines real-world applications and identifies future research directions. Video moment localization holds immense potential to transform industries such as surveillance, e-learning, intelligent content curation, and entertainment by enabling precise and efficient video analysis. By combining foundational principles with state-of-the-art advancements, this book serves as a valuable guide for researchers, practitioners, and students eager to navigate and contribute to this dynamic and evolving field.

It is our hope that this book not only provides a solid foundation for understanding video moment localization but also inspires further innovation and collaboration. The approaches and insights presented here aim to serve as a cornerstone for the development of the next generation of intelligent video retrieval technologies, driving advancements in multimedia analysis and beyond.

Jinan, China — Meng Liu
Jinan, China — Yupeng Hu
Shenzhen, China — Weili Guan
Shenzhen, China — Liqiang Nie

Acknowledgement This book would not have reached completion, or taken its current shape, without the invaluable support and contributions of many colleagues, collaborators, and institutions. It is with great pleasure and heartfelt gratitude that we acknowledge the efforts of all those who have supported and enriched this endeavor.

First and foremost, we extend our sincere thanks to our colleagues who played an essential role in shaping this book. Their active participation in technical discussions, constructive feedback, and insightful suggestions greatly enhanced the quality and depth of the content presented here. We are particularly grateful to those who devoted their time to reviewing and refining early drafts, whose meticulous attention to detail improved the clarity and coherence of the material.

We also wish to express our appreciation to the anonymous reviewers for their thoughtful and constructive feedback during the review process. Their critical evaluations and suggestions have been instrumental in ensuring the rigor and scholarly quality of this work.

Special thanks go to the publisher and the editorial team for their unwavering support and guidance throughout the publishing process. Their expertise, patience, and professionalism have made this journey both smooth and rewarding.

Finally, our deepest gratitude is reserved for our families. Their unconditional love, patience, and encouragement have been a constant source of strength and motivation throughout the process. This book would not have been possible without their unwavering belief in our work and their support during the many hours devoted to its completion.

To everyone who has contributed, directly or indirectly, to the realization of this book, we extend our heartfelt appreciation. Your support and dedication have made this project possible, and we are truly grateful for your contributions.

Competing Interests The authors have no competing interests to declare that are relevant to the content of this manuscript.

Jinan, China — Meng Liu
Jinan, China — Yupeng Hu
Shenzhen, China — Weili Guan
Shenzhen, China — Liqiang Nie
January 2025

Contents

Chapter 1
Introduction

1.1 Background

The rapid growth of digital content and the widespread use of video-sharing platforms have led to an unprecedented increase in video data, including surveillance, educational, entertainment, and social media videos [5]. As a result, efficiently retrieving specific moments within these large-scale, unstructured video datasets has become an essential challenge in the field of multimedia analysis [6].

Unlike static images, videos capture the temporal evolution of objects and actions, making it crucial to understand both the spatial and temporal dimensions of video content. Over the past decade, significant research has focused on action recognition (as illustrated in Fig. 1.1a), where videos are typically classified based on a predefined set of action categories [10]. However, real-world video datasets often consist of long, untrimmed sequences, as seen in applications such as surveillance [12], robotics [1], and autonomous driving [3], which require more sophisticated approaches to both classify and localize actions in time.

To address these challenges, temporal action localization (as illustrated in Fig. 1.1b) has emerged as an important task. Temporal action localization simultaneously classifies and localizes actions within untrimmed video segments [11]. However, these methods are often constrained by predefined action categories, which limits their applicability to the diverse and dynamic range of activities in real-world environments. For example, a query like "The man dribbles the ball towards the basket, then runs to the basket and shoots" involves a complex sequence of events that cannot be easily captured by fixed action categories.

Video moment localization overcomes this limitation by allowing users to query videos using natural language. Instead of relying on predefined categories, users can specify a wide range of activities, making the task more flexible and intuitive. In this context, video moment localization seeks to identify the temporal segment that corresponds to a given natural language query, as illustrated in Fig. 1.1c. This

M. Liu et al., *Intelligent and Efficient Video Moment Localization*,
https://doi.org/10.1007/978-3-031-87588-5_1

Fig. 1.1 Illustration of the action recognition, temporal action localization, and video moment localization (retrieval) tasks. (**a**) The action recognition task aims to classify trimmed videos into different action categories. (**b**) Temporal action localization automatically recognizes the interval of the action occurring in the untrimmed video and classifies the detected action. (**c**) Video moment localization identifies the temporal moment referenced in a natural language query

ability to dynamically retrieve specific video segments based on natural language queries represents a significant advancement in video understanding.

1.2 Related Tasks

Several related tasks in video analysis complement and intersect with video moment localization. These tasks share common methodologies but have distinct objectives, as outlined below.

Temporal Action Localization [7, 9] This task aims to simultaneously classify and localize actions from predefined categories within untrimmed videos. While similar to video moment localization, the primary difference lies in the flexibility of the query. Video moment localization allows for the specification of arbitrary, complex actions through natural language queries, while temporal action localization is constrained to predefined actions. Techniques from temporal action localization, such as proposal generation and temporal relationship modeling, provide valuable foundations for video moment localization.

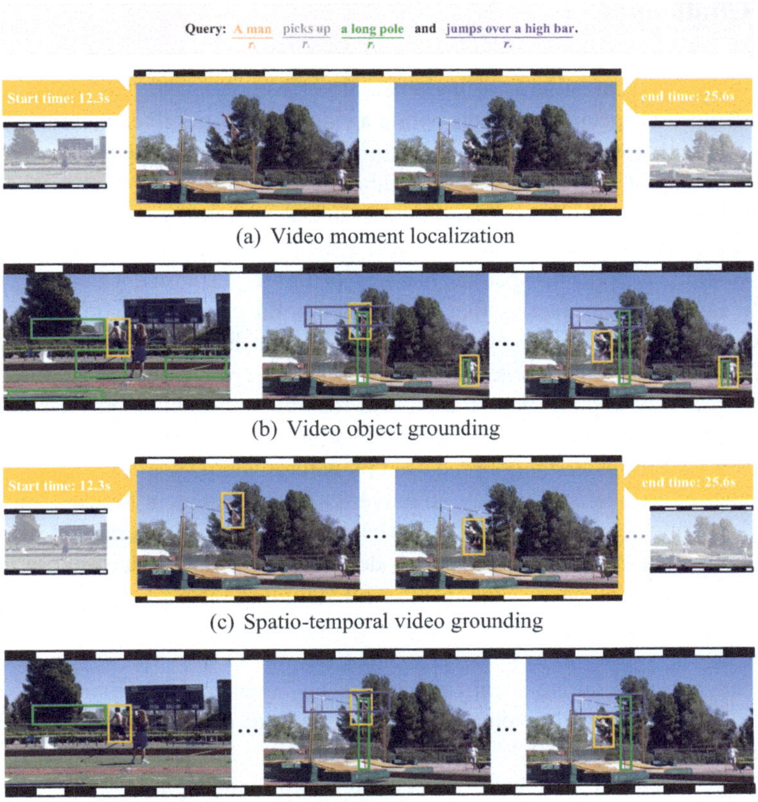

(a) Video moment localization

(b) Video object grounding

(c) Spatio-temporal video grounding

(d) Video semantic role grounding

Fig. 1.2 Illustration of four video grounding tasks: (**a**) Video moment localization, (**b**) Video object grounding, (**c**) Spatio-temporal video grounding, and (**d**) Video semantic role grounding. Various semantic roles are differentiated by color

Video Object Grounding This task addresses the challenge of localizing objects mentioned in a language query [13]. It can be broken down into three categories: (1) Localizing individual objects (Fig. 1.2b), (2) Localizing spatio-temporal tubes of objects (e.g., video re-identification) [2, 4] (Fig. 1.2c), and (3) Resolving ambiguities in object references to identify specific entities in space and time [8] (Fig. 1.2d). While video moment localization is focused on identifying temporal segments described in a query, video object grounding often involves spatial localization or spatio-temporal tracking, making it distinct from the former. Nonetheless, both tasks share technical challenges, such as cross-modal representation learning and query understanding.

1.3 Challenges

Video moment localization is a fundamental yet complex task that involves accurately identifying specific moments within a video based on a given natural language query. This task not only requires determining the precise start and end points of the target segment but also presents several inherent challenges.

A primary difficulty lies in handling the diverse spatial-temporal characteristics and varying durations of video moments. Effective localization demands a deep understanding of both the specific features of the target moment and its broader context. For instance, queries that include temporal constraint words like "first" require comprehension of the overall temporal relationships within the video, adding an additional layer of complexity. This raises the critical question: how can a model effectively capture both the precise features and contextual elements of a video moment, especially when faced with variability in moment durations and the nuanced temporal understanding required by natural language queries?

Another challenge is achieving effective semantic alignment between the query and video content, particularly when queries involve complex temporal keywords such as "first" or "leaving". Existing models often struggle to decompose queries into relevant and irrelevant components while aligning them semantically with the video. The difficulty lies in capturing both the immediate and broader temporal context of the video moment, ensuring that the model uses the key components of the query for accurate localization. The key question here is: how can we dynamically identify and emphasize the most relevant parts of the query to ensure precise moment localization?

Accurately learning moment representations while managing the computational overhead of processing both video and query modalities requires a delicate balance. Specifically, addressing redundant moment candidates and minimizing unnecessary computational costs is crucial. Effective pruning of irrelevant features is essential to capture semantic relevance efficiently without sacrificing localization accuracy. The challenge is: how can we maintain high localization accuracy while optimizing computational efficiency, particularly by pruning redundant moment candidates and aligning semantic information across modalities?

Moreover, localizing moments requires not only understanding intra-modal semantic information (visual or textual) but also capturing inter-modal correlations (consistency and complementarity). While much of the research has focused on intra-modal understanding, there is a pressing need for better collaboration between modalities to improve localization accuracy. The challenge lies in effectively capturing and combining information from both the visual and textual modalities. The question, then, is: how can we design a model that fosters semantic collaboration between modalities, ensuring effective intra-modal understanding and inter-modal alignment for improved moment localization?

Traditional methods of video moment localization are heavily reliant on moment-level annotations, which are both expensive and labor-intensive to acquire. The challenge in weakly-supervised learning lies in localizing moments with only video-level annotations, as the absence of fine-grained temporal annotations makes model training difficult. The question here is: how can we effectively localize video moments in a weakly supervised setting while maintaining performance comparable to fully supervised methods?

As the volume of video data continues to grow, scalability and efficiency become crucial in video moment localization. The challenge is to develop models capable of handling large datasets, processing long video sequences, and ensuring fast moment retrieval. Efficient moment candidate generation, intelligent temporal context modeling, and cross-modal hashing for large-scale retrieval further complicate the task. The key question is: how can we achieve high scalability and efficiency in video moment localization while ensuring effective temporal context modeling and accurate moment candidate generation, particularly with cross-modal hashing techniques for large-scale video datasets?

These challenges highlight the complexity of video moment localization and underscore the need for innovative approaches to address the issues of semantic alignment, contextual understanding, and computational efficiency.

1.4 Our Solutions

To address the challenges in video moment localization, we propose a suite of state-of-the-art methods, each tailored to tackle specific aspects of the problem, including contextual understanding, semantic alignment, efficiency, scalability, and weakly-supervised learning.

Semantic Enhanced Video Moment Localization To overcome the challenge of understanding both the target moment and its surrounding context, we propose the attentive cross-modal retrieval network. This method integrates a memory attention mechanism that prioritizes visual features referenced in the query while incorporating contextual information, thereby enabling enriched moment representations. A cross-modal fusion sub-network is introduced to capture both intra-modality and inter-modality interactions, enhancing the representation of moment-query pairs. This approach effectively addresses the diverse spatial-temporal characteristics of video moments and improves the overall localization accuracy.

Semantic Alignment Video Moment Localization To tackle the challenge of semantic alignment and capturing temporal keywords in queries, we propose the language-temporal attention network. This model decomposes the query into relevant and irrelevant components, dynamically learning word-level attention based on the temporal context of the video. By focusing on the most crucial parts of the query, the network improves semantic alignment, ensuring that the temporal keywords are accurately mapped to the corresponding video moments.

Semantic Pruning Video Moment Localization For challenges related to cross-modal semantic alignment and localization efficiency, we introduce the cross-modal semantic alignment network. This method employs a video encoder to generate and represent moment candidates, paired with a query encoder that captures diverse query intentions. A multi-granularity interaction module explores cross-modal correlations, while a semantic pruning strategy eliminates redundant moment candidates, enhancing computational efficiency and reducing the cost of processing irrelevant moments.

Semantic Collaborative Video Moment Localization To improve intra-modal understanding and foster better inter-modal collaboration, we present the cross-modal semantic perception network. This model incorporates a dual-path representation network for robust intra-modal semantic modeling and a semantic collaborative network for multi-granularity semantic alignment and hierarchical supplementation. This approach ensures effective inter-modal collaboration, addressing the limitations of existing methods that fail to fully exploit necessary semantic complementarity between modalities.

Weakly-Supervised Video Moment Localization In response to the challenge of labor-intensive moment-level annotations, we propose the end-to-end Siamese alignment network for weakly-supervised video moment localization. This network includes a multi-scale Siamese module that bridges the semantic gap between visual and textual modalities. Additionally, a context-aware multiple-instance learning module integrates adjacent context, improving moment-query and video-query alignment. This method allows accurate moment localization with only video-level annotations, significantly reducing the annotation burden.

Efficient Hashing-Based Video Moment Localization To address the challenges of scalability and efficiency, we introduce the deep end-to-end cross-modal hashing network. The model employs a bidirectional temporal convolutional network to generate moment candidates and learn enriched temporal representations. A query encoder captures user intent in a comprehensive manner, and a cross-modal hashing module projects both video and query representations into a shared Hamming space, enabling compact hash code learning. This design facilitates efficient moment retrieval, with pre-computable hash codes that improve scalability and speed.

1.5 Book Structure

The remainder of this book is organized into six chapters, each focusing on a key challenge in video moment localization and presenting cutting-edge solutions.

- Chapter 2 introduces the semantic enhanced video moment localization method. This chapter explores the development of an attentive cross-modal retrieval network, which integrates a memory attention mechanism and a cross-modal fusion sub-network to enhance moment representations and improve contextual understanding.

- Chapter 3 presents the semantic alignment video moment localization approach. Here, we propose a novel language-temporal attention network designed to dynamically decompose queries into relevant and irrelevant components, ensuring precise alignment between textual and visual modalities.
- Chapter 4 delves into semantic pruning video moment localization. This chapter introduces a cross-modal semantic alignment network that leverages a multi-granularity interaction module and a semantic pruning strategy to optimize localization accuracy while minimizing computational overhead.
- Chapter 5 develops the semantic collaborative video moment localization framework. We propose a cross-modal semantic perception network that facilitates hierarchical semantic collaboration, integrating intra-modal semantic modeling and inter-modal complementarity for more accurate localization.
- Chapter 6 explores weakly-supervised video moment localization. In this chapter, we introduce an end-to-end Siamese alignment network, which effectively bridges the semantic gap between visual and textual modalities using weak annotations, supported by a context-aware multiple-instance learning module.
- Chapter 7 introduces the efficient hashing-based video moment localization method. We present a deep end-to-end cross-modal hashing network that ensures scalability and efficiency by leveraging compact hash code learning and offline pre-computation for faster moment retrieval.

Finally, Chap. 8 concludes the book by summarizing the advancements made in video moment localization and identifying potential future research directions to further push the boundaries of this field.

References

1. Burgner-Kahrs, J., Rucker, D.C., Choset, H.: Continuum robots for medical applications: a survey. IEEE Trans. Robot. **31**(6), 1261–1280 (2015)
2. Chen, Z., Ma, L., Luo, W., Wong, K.Y.K.: Weakly-supervised spatio-temporally grounding natural sentence in video. In: Proceedings of the Annual Meeting of the Association for Computational Linguistics, pp. 1884–1894 (2019)
3. Claussmann, L., Revilloud, M., Gruyer, D., Glaser, S.: A review of motion planning for highway autonomous driving. IEEE Trans. Intell. Transp. Syst. **21**(5), 1826–1848 (2020)
4. Li, J., Zhang, S., Huang, T.: Multi-scale temporal cues learning for video person re-identification. IEEE Trans. Image Process. **29**, 4461–4473 (2020)
5. Liu, M., Nie, L., Wang, X., Tian, Q., Chen, B.: Online data organizer: micro-video categorization by structure-guided multimodal dictionary learning. IEEE Trans. Image Process. **28**(3), 1235–1247 (2019)
6. Liu, M., Qu, L., Nie, L., Liu, M., Duan, L., Chen, B.: Iterative local-global collaboration learning towards one-shot video person re-identification. IEEE Trans. Image Process. **29**, 9360–9372 (2020)
7. Qing, Z., Su, H., Gan, W., Wang, D., Wu, W., Wang, X., Qiao, Y., Yan, J., Gao, C., Sang, N.: Temporal context aggregation network for temporal action proposal refinement. In: Proceedings of the IEEE Conference on Computer Vision and Pattern Recognition, pp. 485–494 (2021)

8. Sadhu, A., Chen, K., Nevatia, R.: Video object grounding using semantic roles in language description. In: Proceedings of the IEEE Conference on Computer Vision and Pattern Recognition, pp. 10417–10427 (2020)
9. Su, H., Gan, W., Wu, W., Qiao, Y., Yan, J.: Bsn++: Complementary boundary regressor with scale-balanced relation modeling for temporal action proposal generation. In: Proceedings of the AAAI Conference on Artificial Intelligence, vol. 35, pp. 2602–2610 (2021)
10. Tu, Z., Li, H., Zhang, D., Dauwels, J., Li, B., Yuan, J.: Action-stage emphasized spatiotemporal vlad for video action recognition. IEEE Trans. Image Process. **28**(6), 2799–2812 (2019)
11. Yang, L., Peng, H., Zhang, D., Fu, J., Han, J.: Revisiting anchor mechanisms for temporal action localization. IEEE Trans. Image Process. **29**, 8535–8548 (2020)
12. Zhang, S., Zhu, Y., Roy-Chowdhury, A.K.: Context-aware surveillance video summarization. IEEE Trans. Image Process. **25**(11), 5469–5478 (2016)
13. Zhou, L., Louis, N., Corso, J.J.: Weakly-supervised video object grounding from text by loss weighting and object interaction. In: British Machine Vision Conference, pp. 1–12 (2018)

Chapter 2
Semantic Enhanced Video Moment Localization

2.1 Introduction

In real-world applications such as robotic navigation, autonomous driving, and surveillance, untrimmed videos often contain complex scenes with numerous objects, actions, attributes, and interactions. In these videos, only specific segments are relevant to a given query. For example, in a surveillance video lasting several minutes, a user might be interested in the moment when "a girl in orange first walks by the camera," which occurs between the 24th and 30th seconds, as shown in Fig. 2.1. Therefore, localizing temporal moments of interest within a video is both more useful and more challenging than simply retrieving the entire video. In this chapter, we focus on the task of video moment localization, which aims to precisely identify the start and end points within a video that correspond to a given query. A "desired moment" refers to a query-aware temporal segment whose content matches the given query.[1]

The primary challenges in this task are as follows:

1. **Moment Generation**: Moments within a video vary significantly in terms of duration and spatial-temporal characteristics. This variability makes it difficult to identify relevant moments, let alone estimate their relevance to the query. A common approach for generating moment candidates is to densely sample sliding windows at different temporal scales. However, these methods face limitations, particularly in terms of computational cost and the exponential search space.
2. **Relevance Estimation**: Estimating relevance is a cross-modal retrieval problem, where both visual features of the moment candidates and textual features of the query must be compared in a shared latent space. However, traditional approaches often neglect critical spatial-temporal information in both the moment and the query. For example, in the query "a girl in orange first walks

[1] Note that an entire video may contain multiple moments relevant to the query.

M. Liu et al., *Intelligent and Efficient Video Moment Localization*,
https://doi.org/10.1007/978-3-031-87588-5_2

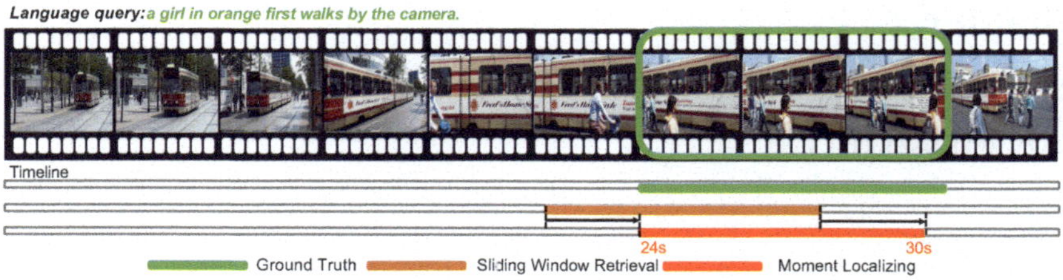

Fig. 2.1 Illustration of video moment localization, where the task is to accurately identify a moment (red bar) with a start time (24th second) and an end time (30th second) based on a given language query. The green bar represents the ground truth moment, the orange bar indicates the result from sliding window-based moment retrieval, and the red bar shows the final localization result achieved by our model

by the camera," the word "first" represents a temporal constraint that requires an understanding of the broader temporal context for correct interpretation.

To tackle these challenges, we propose the Attentive Cross-Modal Retrieval Network (ACRN) for moment retrieval. For moment generation, we introduce a temporal memory attention network that explores the contextual visual features of each moment. Specifically, for each pre-segmented moment, we encode its surrounding context, including preceding and subsequent moments, to capture the continuity of the scene [11]. Inspired by this, we employ a memory network to retain contextual information for each moment. The natural language query is input to an attention network, which adaptively assigns weights to the memory representation, resulting in an enriched moment representation.

Next, we introduce a cross-modal fusion network to enhance the moment-query representation. This network captures both inter-modal and intra-modal interactions. The inter-modal interactions explicitly model relationships between visual and textual embeddings, while the intra-modal interactions explore relationships within each modality. Finally, we use a boundary regression model to predict relevance scores and moment offsets, ensuring accurate localization of the desired moment.

We evaluate our approach on two benchmark datasets, DiDeMo and TACoS, and demonstrate significant improvements over existing methods. Furthermore, we have made the code publicly available to support further research in this area.[2]

2.2 Related Work

This work is related to video retrieval and temporal action localization.

[2] https://sigir2018.wixsite.com/acrn.

Video Retrieval Video retrieval aims to retrieve relevant videos or specific video segments based on a language query [1, 7, 26]. Recent approaches to video retrieval, much like image-language embedding models [8, 24], leverage deep video-language embeddings. Early methods, such as that proposed by Lin et al. [16], focused on aligning visual concepts in videos with semantic graphs generated by parsing sentence descriptions. Additionally, Alayrac et al. [2], Bojanowski et al. [4], and Sener et al. [20] addressed video-text alignment by assigning temporal intervals to sentences within the context of the video, aiming to capture the temporal ordering of events.

In a separate direction, Hendricks et al. [3] introduced a joint video-language model designed to retrieve moments based on textual queries. However, their model faces challenges in handling background noise, as it only verifies the presence of corresponding moments without considering their specific temporal boundaries. While such models can densely sample video moments at various temporal scales for retrieval, this approach can be computationally expensive and suffers from an enlarged search space, complicating the matching process. To mitigate this issue, we draw inspiration from the success of temporal boundary adjustment in object localization [18], where regression parameters are used to predict the start and end times of relevant video moments.

Temporal Action Localization Another important area of research is the temporal localization of actions in untrimmed videos. Gaidon et al. [9, 10] were among the first to introduce this challenge, initially focusing on limited action categories such as "drinking" or "smoking." As more complex action categories became available, new methods emerged for localizing activities. For example, Shou et al. [21] introduced a 3D Convolutional Neural Network (CNN) that outperforms Recurrent Neural Network (RNN)-based methods by effectively capturing spatio-temporal features. Singh et al. [23] proposed a multi-stream bi-directional RNN to capture fine-grained action detection, while Gao et al. [12] introduced a temporal unit regression network designed to jointly predict action proposals and refine temporal boundaries.

More recently, Gao et al. [11] incorporated natural language queries for localizing activities, focusing on the surrounding context without explicitly considering the semantics of the input query. This approach can lead to the selection of irrelevant or misleading video moments, as the model does not fully analyze the content and meaning of the query itself.

Although temporal action localization and video moment retrieval share common challenges, such as understanding temporal boundaries, video moment retrieval introduces additional complexities. These include the need to capture intricate, query-specific semantics, which is a central focus of our proposed method.

2.3 Methodology

As illustrated in Fig. 2.2, our proposed ACRN model consists of the following components: (1) Memory Attention Network: This component utilizes contextual weighting to enhance the visual embedding of each moment, making the representation more informative by focusing on relevant segments within the video. (2) Cross-Modal Fusion Network: This network explores both intra-modal and inter-modal feature interactions to generate effective moment-query representations, capturing relationships both within and across the visual and textual modalities. (3) Regression Network: This component estimates the relevance scores and predicts the location offsets of the retrieved moments, refining the temporal boundaries for more accurate moment localization.

2.3.1 *Problem Formulation*

Let v and q denote a video and a query, respectively. The video v is represented as a sequence of frames $v = f_t$, where f_t represents a frame at time index $t \in \{0, \cdots, \tau\}$. The query q is associated with a temporal annotation (t_s, t_e), where t_s and t_e represent the start and end points of the corresponding moment. For simplicity, we assume that a description query corresponds to a single relevant moment within the video, though multiple moments may be relevant.

Our task is to accurately identify a golden moment $c = \{f_t\}_{t=\tau_s}^{\tau_t}$ within the video that matches the given query q, such that the temporal boundaries (τ_s, τ_e) align with

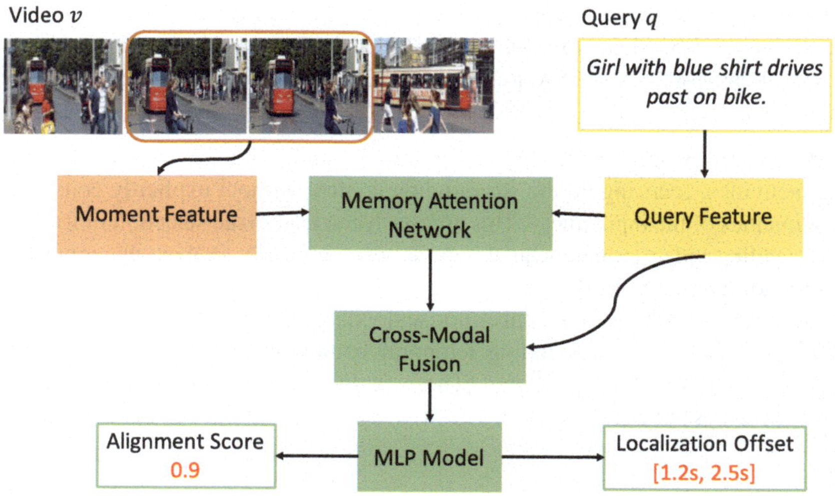

Fig. 2.2 An illustration of our proposed ACRN model

(t_s, t_e). To achieve this, we pre-segment the video v into a set of moment candidates $C = \{c_i\}_{i=1}^{M}$ using multi-scale temporal sliding windows, where M is the number of generated moments.[3]

For the given query q and a moment candidate c that overlaps with the golden moment, we align the moment-query pair as a positive training sample. Since positive candidates may overlap with the golden moment at different scales due to the segmentation strategy, each positive moment-query pair (c, q) is paired with a temporal offset, defined as $(t_s - \tau_s, t_e - \tau_e)$. The process of constructing training data is described in Sect. 2.4.1.

Thus, the problem of moment retrieval can be formally defined as:

Input: A set of moment candidates C and a given query q.

Output: A ranking model that maps each moment-query pair (c, q) to a relevance score and estimates the location offsets for accurately localizing the golden moment.

2.3.2 *Memory Attention Network*

To estimate the matching score between each moment candidate and the given query, a straightforward approach is to project the visual embeddings of the moment candidates and the textual embedding of the query into a shared latent space. This is followed by using a well-designed similarity function to calculate their relevance and selecting the moment with the highest score as the retrieval result. Formally, this process can be summarized as follows:

$$\begin{cases} \widehat{\mathbf{x}}_c = f_\Theta(\mathbf{x}_c), \\ \widehat{\mathbf{q}} = f_\Theta(\mathbf{q}), \\ c^* = \arg\max_{c \in C} g(\widehat{\mathbf{x}}_c, \widehat{\mathbf{q}}), \end{cases} \tag{2.1}$$

where $\mathbf{x}_c \in \mathbb{R}^{D_1}$ and $\mathbf{q} \in \mathbb{R}^{D_2}$ denote the embeddings of the moment c and the input query q, respectively. The function $f_\Theta(\cdot)$ projects $\mathbf{x}_c$ and $\mathbf{q}$ into a common latent space, resulting in $\widehat{\mathbf{x}}_c$ and $\widehat{\mathbf{q}}$. The function g then calculates the similarity between these embeddings.

Although this approach is feasible, it only considers the current moment candidate, neglecting the surrounding spatial-temporal context. This omission may lead to information loss and suboptimal performance. For example, in the query "a girl in orange first walked by the camera," the word "first" introduces a temporal constraint that requires a broader temporal context to be properly understood.

[3] The generation of moment candidates, the visual embedding of each moment, and the textual embedding of the given query are detailed in Sect. 2.4.1.

Prior work [11] has shown that pre-context and post-context moments provide crucial temporal information and help determine the relative temporal position of a moment within a video. Inspired by these observations, we leverage context information to complement the current moment.

Let the context moments of each video moment $c \in C$ be represented as $\mathcal{N}_c = \{c_j\}$, where $j \in [-n_c, n_c]$ and n_c denotes the context shift boundary.[4] We use $j > 0$, $j = 0$, and $j < 0$ to indicate post-context, current, and pre-context moments, respectively. The embedding of the central moment c is denoted as $\mathbf{x}_c$, and the embeddings of its context moments are denoted as $\mathbf{x}_{c_j}$. How these contextual embeddings are integrated is crucial for strengthening the representation of the current moment. A simple approach adopted in [12] is to employ average pooling over the context embeddings to capture interactions between the current moment and its context. The enhanced representation of the current moment is given by:

$$\widehat{\mathbf{x}}_c = \frac{1}{|\mathcal{N}_c|} \sum_{c_j \in \mathcal{N}_c} \mathbf{x}_{c_j}. \tag{2.2}$$

While average pooling fuses all contextual embeddings into a single representation, it is limited in capturing consistent information and complex interactions among context moments. Specifically, average pooling assumes that the moments are linearly independent and contribute equally to the final relevance estimation, which fails to differentiate the importance of each moment and may not effectively eliminate noisy or irrelevant features.

To overcome this limitation, we explicitly model the varying importance of each context moment by assigning an attention weight to each context embedding. Our memory attention mechanism, illustrated in Fig. 2.3, accounts for two factors that contribute to the attention scores: (1) the relevance between the moment and the query, and (2) the temporal relationships between consecutive moments.

Given the representation vector of a context moment $\mathbf{x}_{c_j} \in \mathbb{R}^{D_1}$ and the representation of the given query $\mathbf{q} \in \mathbb{R}^{D_2}$, we use a one-layer network to estimate the attention score α_{c_j}, which reflects the consistency between the moment and the query. Additionally, for each moment in the context, we incorporate the representations of prior moments to capture temporal information and better model the importance weights. Formally, our memory attention mechanism is presented as follows:

$$\begin{cases} e(c_j, \mathbf{q}) = \sigma(\sum_{i=-n_c}^{j} \mathbf{W}_c \mathbf{x}_{c_i} + \mathbf{b}_c)^T \cdot \sigma(\mathbf{W}_q \mathbf{q} + \mathbf{b}_q), \\ \alpha_{c_j} = \frac{e(c_j, \mathbf{q})}{\sum_{k=-n_c}^{n_c} e(c_k, \mathbf{q})}, \; j \in [-n_c, n_c], \end{cases} \tag{2.3}$$

[4] For example, $n_c = 1$ implies that there is one pre-context and one post-context moment. The process of generating context moments is described in Sect. 2.4.1.

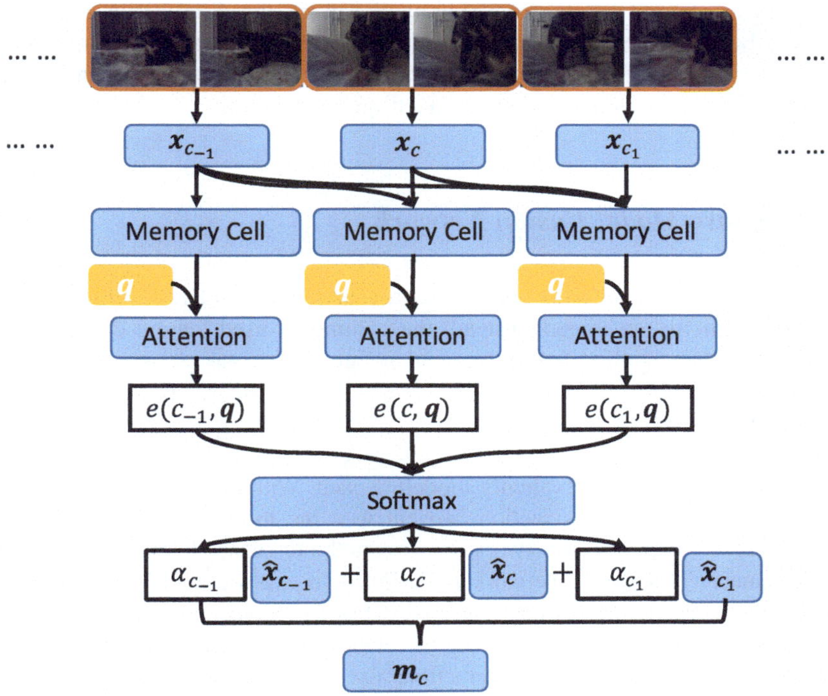

Fig. 2.3 An illustration of our proposed memory attention model

where $\mathbf{W}_c \in \mathbb{R}^{D \times D_1}$ and $\mathbf{W}_q \in \mathbb{R}^{D \times D_2}$ are used to transform the query and video embeddings to a common underlying embedding space. $\mathbf{b}_c \in \mathbb{R}^D$ and $\mathbf{b}_q \in \mathbb{R}^D$ denote the trainable bias vector for the moment and the query, respectively; and $\sigma(\cdot)$ is the tanh activation function to restrict the attention weight to be in (0,1).

With the attention weight α_{c_j}, the fused feature $\mathbf{m}_c$ is computed as follows:

$$\begin{cases} \widehat{\mathbf{x}}_{c_j} = \mathbf{W}_c \mathbf{x}_{c_j} + \mathbf{b}_c, \\ \mathbf{m}_c = \sum_{j \in [-n_c, n_c]} \alpha_{c_j} \widehat{\mathbf{x}}_{c_j}, \end{cases} \tag{2.4}$$

where $\mathbf{m}_c \in \mathbb{R}^D$ represents the enhanced embedding of the current moment c, attended by the input query. The embedding matrix $\mathbf{W}_c$ and bias vector $\mathbf{b}_c$ are the same as in Eq. (2.3). Thus, the memory attention network leverages the context weights of each moment's varying importance to enhance the moment representations.

Finally, we obtain the embedding query feature:

$$\widehat{\mathbf{q}} = \mathbf{W}_q \mathbf{q} + \mathbf{b}_q, \tag{2.5}$$

where $\mathbf{W}_q$ and $\mathbf{b}_q$ are the query embedding matrix and bias vector from Eq. (2.3).

Thus, the memory attention network effectively incorporates varying importance weights for each moment's context to enhance the representation of the current moment, resulting in more accurate moment-query matching.

2.3.3 Cross-Modal Fusion Network

In prior multimodal research, intra-modality and inter-modality dynamics have not been explicitly leveraged; instead, the commonly-used feature concatenation approach is often applied for multimodal feature fusion (as shown in Fig. 2.4a). However, this approach does not efficiently model the relationships between different modalities.

In this chapter, we introduce a fusion sub-network that disentangles unimodal and bimodal dynamics by explicitly modeling each. With the attentive embeddings established in the previous section, we enhance the moment representation and design a cross-modal fusion network to estimate the relevance between the moment and the query. This fusion network is designed to explore both intra-modal and inter-modal feature interactions: intra-modal interactions are implemented via tensor fusion, which explicitly models the relationships between visual and textual embeddings, while inter-modal interactions use concatenation to retain modality-specific information. Ultimately, these intra-modal and inter-modal embeddings are concatenated to create a fused moment-query representation.

As depicted in Fig. 2.4b, the cross-modal fusion network consists of two key components: mean pooling and tensor fusion. Since high-dimensional vectors can significantly increase computational complexity during tensor fusion, we introduce a mean pooling layer before performing tensor fusion. Specifically, given a D-dimensional moment embedding $\mathbf{m}_c$ and a D-dimensional query embedding $\widehat{\mathbf{q}}$ from the preceding memory attention network, we use mean pooling to reduce dimensionality while retaining essential features. The mean pooling operation can be seen as applying a linear filter with window size n to each input embedding, where each output value is the mean of the corresponding window.

After applying mean pooling to $\mathbf{m}_c$ and $\widehat{\mathbf{q}}$, we obtain dimension-reduced, high-level representations $\widetilde{\mathbf{m}}_c$ and $\widetilde{\mathbf{q}}$ for the moment and the query, respectively. These embeddings are then input into the tensor fusion model. Tensor fusion can be viewed as a differentiable outer product between the visual representation $\widetilde{\mathbf{m}}_c$ and the query representation $\widetilde{\mathbf{q}}$:

$$\mathbf{f}_{cq} = \begin{bmatrix} \widetilde{\mathbf{m}}_c \\ 1 \end{bmatrix} \otimes \begin{bmatrix} \widetilde{\mathbf{q}} \\ 1 \end{bmatrix} = [\widetilde{\mathbf{m}}_c, \widetilde{\mathbf{m}}_c \otimes \widetilde{\mathbf{q}}, \widetilde{\mathbf{q}}, 1], \tag{2.6}$$

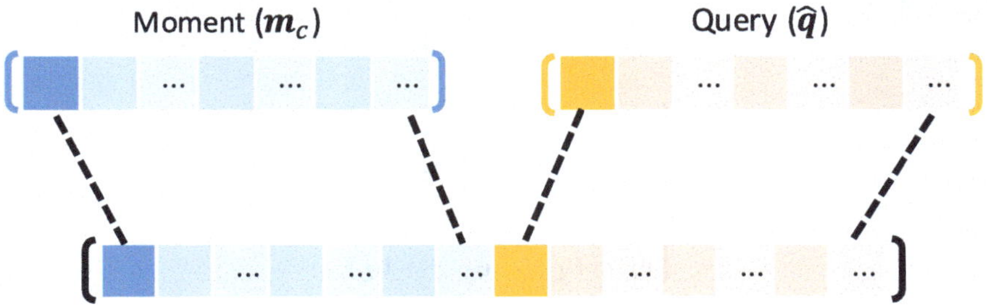

(a)

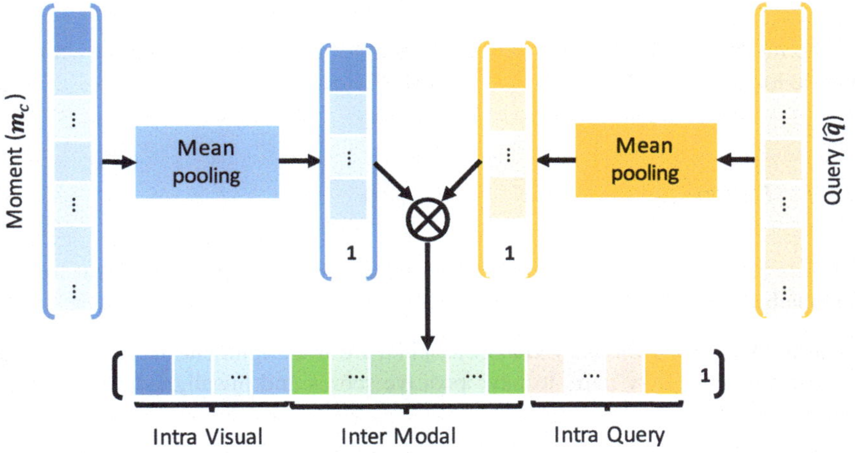

(b)

Fig. 2.4 Illustration of the commonly used early fusion technique compared to our proposed cross-modal feature fusion model. The top image represents early fusion using multimodal concatenation, while the bottom image depicts our cross-modal feature fusion model, which incorporates both intra-modal and inter-modal feature interactions. (**a**) Early fusion. (**b**) Cross-modal fusion

where $\otimes$ represents the outer product between vectors, and $\mathbf{f}_{cq}$ encompasses all possible combinations of the unimodal embeddings within three semantically distinct subregions. The two subregions $\tilde{\mathbf{m}}_c$ and $\tilde{\mathbf{q}}$ capture unimodal interactions, while $\tilde{\mathbf{m}}_c \otimes \tilde{\mathbf{q}}$ captures bimodal interactions in tensor fusion.

Above the tensor fusion sub-network $\mathbf{f}_{cq}$, we use a multi-layer perceptron (MLP) to compute the matching score of the moment-query pair (c, q) as well as the localization offset between the moment candidate and the ground truth moment $(t_s - \tau_s, t_e - \tau_e)$. The hidden layers of the MLP are defined as follows:[5]

$$\begin{cases} \mathbf{e}_1 = \theta_1(\mathbf{W}_1\mathbf{f}_{cq} + \mathbf{b}_1), \\ \mathbf{e}_2 = \theta_2(\mathbf{W}_2\mathbf{e}_1 + \mathbf{b}_2), \\ \dots \\ \mathbf{e}_L = \theta_L(\mathbf{W}_L\mathbf{e}_{L-1} + \mathbf{b}_L), \end{cases} \tag{2.7}$$

where $\mathbf{W}_l$, $\mathbf{b}_l$, θ_l, and $\mathbf{e}_l$ represent the weight matrix, bias vector, activation function, and output vector of the l-th hidden layer, respectively. We use ReLU as the activation function for each hidden layer.

The final output vector $\mathbf{e}_L = [s_{cq}, \delta_s, \delta_e] \in \mathbb{R}^3$ contains the matching score s_{cq} and the localization offsets $\delta_s = t_s - \tau_s$ and $\delta_e = t_e - \tau_e$.

2.3.4 *Learning*

The loss function of our proposed model consists of two parts: alignment loss for the matching score and localization regression loss for the offsets.

Alignment Loss Inspired by Gao et al. [11], we use alignment loss to encourage aligned moment-query pairs to have positive scores and misaligned pairs to have negative scores:

$$L_{align} = \alpha_1 \sum_{(c,q)\in\mathcal{P}} \log(1 + \exp(-s_{cq})) + \alpha_2 \sum_{(c,q)\in\mathcal{N}} \log(1 + \exp(s_{cq})), \tag{2.8}$$

where $\mathcal{P}$ is the set of aligned moment-query pairs (positive pairs), and $\mathcal{N}$ is the set of misaligned moment-query pairs (negative pairs). The hyperparameters α_1 and α_2 control the balance between positive and negative pairs.

Localization Regression Loss Since multi-scale temporal sliding windows are used to segment videos, different moment candidates have varying durations. For each moment-query pair, it is essential to not only determine the relevance of the moment to the query but also estimate the localization offsets with respect to the ground truth. We adopt the moment boundary adjustment strategy presented in [12].

We define the offsets for the start and end points as follows:

$$\begin{cases} \delta_s^* = t_s - \tau_s, \\ \delta_e^* = t_e - \tau_e, \end{cases} \tag{2.9}$$

[5] In our experiments, the MLP has two layers.

where (t_s, t_e) are the start and end points of the ground truth, and (τ_s, τ_e) are the start and end points of a candidate moment in $\mathcal{P}$. We use $\delta^* = [\delta_s^*, \delta_e^*]$ to denote the ground truth localization offsets.

Based on the ground truth offsets, we adaptively adjust the alignment of the current moments to match the exact temporal duration using the following localization regression loss:

$$L_{loc} = \sum_{(c,q)\in\mathcal{P}} [R(\delta_s^* - \delta_s) + R(\delta_e^* - \delta_e)], \tag{2.10}$$

where $\mathcal{P}$ is the set of positive moment-query pairs, and R is the L_1 norm function.

Overall Objective The overall optimization objective combines the alignment loss and localization regression loss:

$$L = L_{align} + \lambda L_{loc}, \tag{2.11}$$

where λ is a hyper-parameter that balances the two losses.

2.4 Experiments

We first evaluated the effectiveness of our proposed model on two video moment localization datasets: TACoS and DiDeMo. We then investigated the impact of our well-designed attention network on localization performance.

2.4.1 Experimental Settings

2.4.1.1 Datasets

We evaluated our proposed ACRN model on two benchmark datasets: TACoS and DiDeMo.

TACoS [11] It is constructed based on the MPII Composite dataset [19] and contains 127 videos. Each video is annotated with both fine-grained activity labels and temporal information (i.e., start and end points), along with natural language descriptions for each activity. The dataset was used by Gao et al. [11] for temporal activity localization and is commonly referred to as TACoS.[6]

The training videos are sampled using multi-scale temporal sliding windows of sizes [64, 128, 256, 512] frames with an 80% overlap. Testing samples are

[6] https://github.com/jiyanggao/TALL.

Table 2.1 Summary of TACoS and DiDeMo datasets

Dataset	# Videos	# Queries	# Moments	Domain	Video Source
TACoS	100	14,229	2326	Cooking	Lab Kitchen
DiDeMo	10,464	40,543	26,892	Open	Flickr

coarsely sampled using sliding windows with sizes [128, 256] frames. A sliding window moment c with temporal annotation (τ_s, τ_e) and a query description q with temporal annotation (t_s, t_e) are aligned as a training pair if they satisfy the following conditions: (1) the Intersection over Union (IoU) is larger than 0.5; (2) the non-intersection over Length (nIoL) is smaller than 0.15; and (3) one sliding window moment can be aligned with only one query description.

The dataset contains 75 training videos, 25 testing videos, and 26,963 training moment-query pairs satisfying these conditions. For feature extraction, we used C3D [25] as the moment-level visual encoder and Skip-thoughts [6, 15] as the query description embedding extractor. Consequently, the visual and textual embeddings are of dimensions 4096 and 4800, respectively.

DiDeMo [3] It is a dataset specifically designed for language-based moment retrieval, called Distinct Describable Moments (DiDeMo).[7] It consists of 10,464 personal videos, each lasting between 25–30 seconds, along with 26,892 video moments and 40,543 localized descriptions. The descriptions refer to expressions used to describe specific moments within a video. A verification step ensures the descriptions align with a single moment within the video.

In DiDeMo, each video is divided into six five-second moments, represented by a 6×4096 feature matrix where each column is a 4096-dimensional VGG [22] feature of one moment. For language features, the dataset employs 300-dimensional GloVe [17] word embeddings to represent each word. A summary of these two datasets is provided in Table 2.1.

The experimental results reported in this chapter are based on the datasets mentioned above. In the following experiments, we set the context moment number n_c to 1. The context window length is set to 128 frames for the TACoS dataset and 5 seconds for the DiDeMo dataset. The experiments were carried out using TensorFlow, with the Adam optimizer for training. The model was trained on a server equipped with 16 Tesla K80 GPUs.

2.4.1.2 Evaluation Protocols

To thoroughly evaluate the performance of our proposed model and the baselines, we adopted the "R@n, IoU=m" metric, initially proposed by Hu et al. [13]. Specifically, given a query, this metric calculates the percentage of the top-n

[7] https://github.com/LisaAnne/LocalizingMoments.

retrieved results that have an IoU greater than or equal to m with the ground truth moment. In our work, we denote this metric as $R(n, m)$.

Since this metric is computed at the individual query level, the overall performance is determined by averaging the recall values over all queries, defined as follows:

$$R(n, m) = \frac{1}{N_q} \sum_{i=1}^{N_q} r(n, m, q_i), \tag{2.12}$$

where $r(n, m, q_i)$ represents the recall value for a given query q_i, and N_q is the total number of queries. $R(n, m)$ is the final averaged performance across all queries, indicating the ability of the model to accurately localize the moments relevant to the queries [5, 27].

2.4.1.3 Baselines

To demonstrate the effectiveness of our proposed ACRN model, we compared it with several state-of-the-art baselines:

- **TALL** [11]: This is a cross-modal temporal regression localizer that jointly captures the interaction between the query description and video moments. It provides alignment scores as well as boundary regression for the temporal localization of moment candidates.
- **MCN** [3]: This method is designed for moment-query retrieval, which focuses on leveraging both local and global moment features to enhance expressiveness and improve moment retrieval performance.
- **VSA-RNN** [11]: This method is a variant of the Deep Visual-Semantic Alignment (DVSA) model [14]. It projects local visual features and textual features (encoded by an LSTM) into a common embedding space, and then estimates the matching score between each moment candidate and the query, as described in Eq. (2.1).
- **VSA-STV** [11]: Similar to VSA-RNN, this model uses Skip-Thought vectors [15] instead of RNNs for encoding the query description. The Skip-Thought encoder produces a 4800-dimensional vector, which is linearly transformed to 1000 dimensions for similarity calculation. The visual encoder used is the same as that in VSA-RNN.

Note that VSA-RNN and VSA-STV are two baseline models described in [11], but the source codes and parameters were not made publicly available by the authors. Therefore, we implemented these two models independently and made extensive efforts to tune their parameters for optimal performance. In our implementation, each word was represented with 500-dimensional dense word embeddings, specifically using Glove [17], for training the VSA-RNN. The hidden state size of the LSTM was set to 1024, and the output size was 1000. Video moments were processed by a visual encoder and linearly transformed to 1000 dimensions, which served as the moment-level embeddings. To train these models, we employed hinge loss, defined as follows:

$$L = \sum_{k}[\sum_{l} \max(0, s_{kl} - s_{kk} + 1) + \sum_{l} \max(0, s_{lk} - s_{kk} + 1)], \tag{2.13}$$

where k is the index of moment candidates, l is the index of query descriptions, s_{kk} denotes the cosine score of an aligned moment-query pair, and s_{kl} or s_{lk} denotes the cosine score of the misaligned moment-query pair.

2.4.2 Performance Comparison

Table 2.2 displays the performance comparison for $R(n, m)$ on TACoS. We make the following observations:

- VSA-STV and VSA-RNN achieve poor performance due to their inability to leverage context information from moment candidates. This limits their capacity to exploit spatial-temporal cues, emphasizing the importance of modeling context in moment retrieval.
- MCN considers the surrounding context but employs average pooling of all context representations, thereby treating each context moment with equal importance. This approach overlooks the adaptive importance of individual context moments, potentially introducing noisy features and resulting in negative transfer. This is reflected in MCN's unstable performance across datasets, highlighting the need for adaptive attention weights for each context moment.
- TALL outperforms MCN, VSA-STV, and VSA-RNN in the moment retrieval task. This improvement is due to TALL's ability to capture cross-modal interactions between visual and textual information, thereby enhancing the expressiveness of the moment-query pairs.
- ACRN achieves the best overall performance, significantly surpassing all baselines. Its consistent improvements over TALL and MCN validate the effective-

Table 2.2 Performance comparison between our proposed ACRN model and state-of-the-art baselines on the TACoS dataset (p-value*: p-value calculated for $R(1, 0.5)$)

Method	R@1 IoU=0.5	R@1 IoU=0.3	R@1 IoU=0.1	R@5 IoU=0.5	R@5 IoU=0.3	R@5 IoU=0.1	p-value*
MCN	1.25%	1.64%	3.11%	1.25%	2.03%	3.11%	3.62E-10
VSA-STV	8.84%	13.59%	17.58%	16.41%	26.40%	35.86%	2.16E-06
VSA-RNN	9.96%	16.16%	20.92%	18.32%	29.19%	40.66%	1.82E-05
TALL	11.22%	15.50%	20.21%	23.46%	31.37%	44.40%	5.71E-05
ACRN	**14.62%**	**19.52%**	**24.22%**	**24.88%**	**34.97%**	**47.42%**	–

The best experimental results are highlighted in bold

Table 2.3 Performance comparison between our proposed ACRN model and state-of-the-art baselines on the DiDeMo dataset (p-value*: p-value calculated for $R(1, 0.5)$)

Method	R@1 IoU=0.5	R@1 IoU=0.7	R@1 IoU=0.9	R@5 IoU=0.5	R@5 IoU=0.7	R@5 IoU=0.9	p-value*
MCN	23.33%	15.37%	15.32%	41.03%	20.37%	19.77%	6.14E-09
VSA-STV	25.38%	14.49%	14.39%	68.56%	26.92%	24.24%	1.98E-03
VSA-RNN	24.94%	14.52%	14.44%	68.39%	26.10%	23.95%	3.31E-06
TALL	26.45%	15.36%	15.31%	68.78%	28.43%	26.15%	2.32E-02
ACRN	**27.44%**	**16.65%**	**16.53%**	**69.43%**	**29.45%**	**26.82%**	–

The best experimental results are highlighted in bold

ness of incorporating context memory and employing an attention mechanism to adaptively determine the importance of each context moment.

We also evaluated our proposed ACRN model on the DiDeMo dataset, comparing it to the baseline methods in terms of IoU values of {0.5, 0.7, 0.9} and R@{1, 5}. Since the positive moment-query pairs in this dataset are well aligned without location offsets, we only used the alignment loss to train ACRN and TALL for localization. The results, presented in Table 2.3, show consistent trends with those on TACoS: ACRN significantly outperforms the non-attention models (TALL and MCN) as well as the non-context models (VSA-RNN and VSA-STV).

Furthermore, a significance test was conducted to compare ACRN against each baseline. All p-values were found to be significantly smaller than 0.05, indicating that the improvements achieved by our model are statistically significant.

2.4.3 Ablation Study

We experimented with variants of our model to verify the effectiveness of the memory attention and cross-modal fusion networks:

- **ACRN-a**: We utilized average pooling (Eq. (2.2)) instead of our memory attention network to integrate context embeddings.
- **ACRN-m**: We eliminated the memory component of our memory attention network (Eq. (2.3)). The attention value for each context is only influenced by itself and the query, without considering the broader context.
- **ACRN-c**: We utilized early fusion (i.e., multimodal feature concatenation) in place of the cross-modal fusion model (Eq. (2.6)).

We evaluated these model variants on the TACoS and DiDeMo datasets, and the results are presented in Fig. 2.5.

By analyzing Fig. 2.5, we have the following observations:

- As shown in Fig. 2.5a–d, ACRN significantly outperforms ACRN-a on both the DiDeMo and TACoS datasets. This suggests that average pooling is insufficient to capture the consistency and complex interactions among the moment contexts.

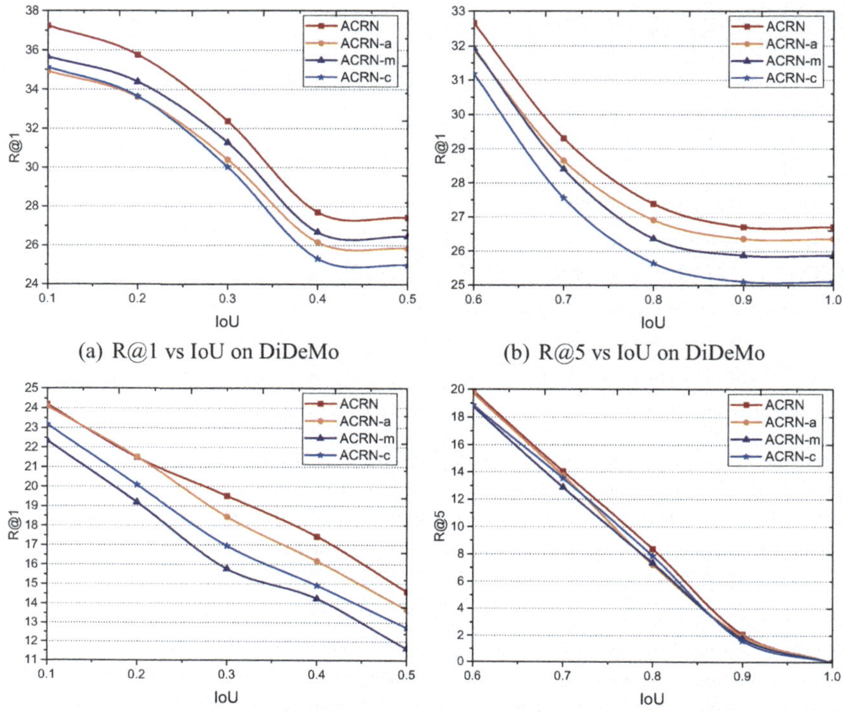

(a) R@1 vs IoU on DiDeMo (b) R@5 vs IoU on DiDeMo

(c) R@1 vs IoU on TACoS (d) R@5 vs IoU on TACoS

Fig. 2.5 Performance comparison among our model variants on the TACoS and DiDeMo datasets. (**a**) R@1 vs. IoU $\in \{0.1, 0.2, 0.3, 0.4, 0.5\}$ on DiDeMo; (**b**) R@5 vs. IoU $\in \{0.6, 0.7, 0.8, 0.9, 1.0\}$ on DiDeMo; (**c**) R@1 vs. IoU $\in \{0.1, 0.2, 0.3, 0.4, 0.5\}$ on TACoS; (**d**) R@5 vs. IoU $\in \{0.6, 0.7, 0.8, 0.9, 1.0\}$ on TACoS

Average pooling assumes that context moments contribute equally, failing to highlight the importance of specific moments or eliminate noisy features. The improvement observed with ACRN demonstrates the effectiveness of the attention mechanism.

- Removing the memory component (ACRN-m) hurts the expressiveness of moment representation, degrading retrieval performance. ACRN-m assumes the representation of each moment is independent of its surrounding context, failing to leverage the spatial-temporal information in the context. By employing memory attention, ACRN enriches moment representations, improving performance.
- ACRN consistently outperforms ACRN-c on both datasets, highlighting the importance of modality interaction. Concatenation only models intra-modal interactions, limiting the expressiveness of moment-query representations. The cross-modal fusion network in ACRN effectively captures both intra-modal and inter-modal interactions, resulting in richer moment-query representations.

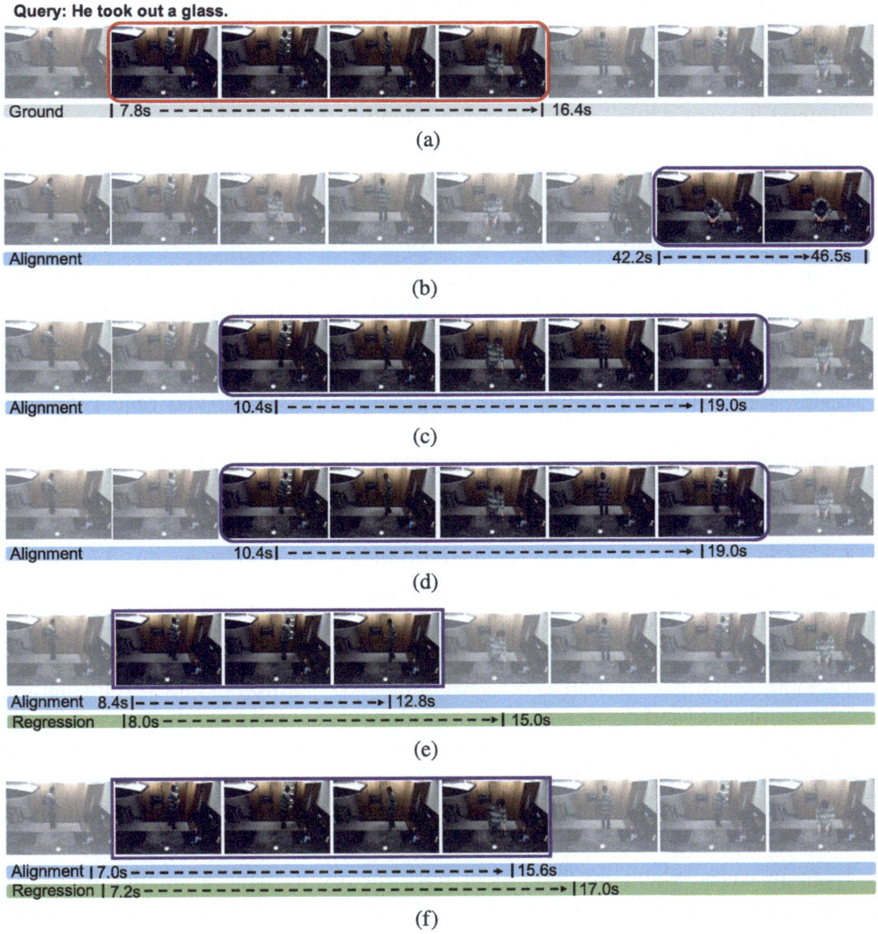

Fig. 2.6 Moment retrieval results on the TACoS dataset. All of the above figures are the R@1 results. The gray, blue and green bars denote the time line of the ground truth, alignment result and regression fine tune result, respectively. (**a**) The golden moment of the moment retrieval. (**b**) The moment retrieval result of the MCN. (**c**) The moment retrieval result of the VSA-STV. (**d**) The moment retrieval result of the VSA-RNN. (**e**) The moment retrieval result of the TALL. (**f**) The moment retrieval result of the ACRN

2.4.4 Qualitative Results

To gain deeper insights into our proposed ACRN model, we presented an example of moment retrieval. The video shown in Fig. 2.6 depicts a complex cooking scene: a man takes a glass from the cupboard, places it on the countertop, goes back to the cupboard to take a second glass, and later cracks an egg over the glass to drain the egg white. We used the description "He took out a glass" from the dataset as the query and apply the models to retrieve the relevant moment. From the results presented in Fig. 2.6, we observed the following:

- As illustrated in Fig. 2.6b, MCN retrieves a moment involving "The man drained the egg white" rather than the intended moment "He took out a glass". Although MCN considers both local and global moment features, it combines all background contexts as global features, which can lead to misrepresentation. Since many moments in the video involve the "egg cracking and draining" scene, the global visual embedding fails to effectively represent the desired scene.
- Both VSA-STV and VSA-RNN return a moment containing two sub-scenes: "He took out a glass" and "He took a second glass", as depicted in Figs. 2.6c, d. Since these models only consider the current moment without leveraging temporal context, they cannot accurately determine the relative order of actions. As a result, they return all frames containing the action "took" and object "glass". This limitation emphasizes the importance of incorporating spatial-temporal information from the surrounding context.
- TALL produces more accurate alignment compared to MCN, VSA-STV, and VSA-RNN, as shown in Fig. 2.6e. This highlights the significance of cross-modal fusion in enriching moment-query representations by capturing intra-modal and inter-modal feature interactions.
- Our ACRN model outperforms all state-of-the-art baselines. As seen in Fig. 2.6f, our alignment has a larger IoU with the ground truth moment. Moreover, ACRN provides better results than TALL. Although TALL utilizes context information, it pools the pre- and post-contexts into a single vector and concatenates it with the current moment. This approach ignores complex interactions among contexts and cannot accurately assess the importance of each moment, missing essential cues. The performance of ACRN validates the effectiveness of our attentive moment retrieval network.
- Even in cases where the alignment retrieval results have a smaller IoU with the ground truth moment, ACRN and TALL can correct the alignment using their regression components, leading to more precise results. This underlines the effectiveness of location offset regression. Furthermore, ACRN demonstrates better offset performance than TALL, indicating that ACRN generates a more refined moment-query representation.

2.5 Summary

In this chapter, we present an attentive cross-modal retrieval framework designed to retrieve specific moments from long videos based on a given query. To effectively align the moment candidates with the query, we introduce a memory attention model that dynamically computes visual attention by considering both the query and its contextual information. Furthermore, we employ a cross-modal fusion sub-network to better integrate and align the cross-modal information between the moment and the query.

We evaluate our approach through extensive experiments on two public datasets, TACoS and DiDeMo. The results demonstrate that our model significantly outperforms state-of-the-art baselines, highlighting the effectiveness of our proposed methods for moment-query alignment in the video moment localization task.

References

1. Al Kabary, I., Schuldt, H.: Enhancing sketch-based sport video retrieval by suggesting relevant motion paths. In: Proceedings of the International ACM SIGIR Conference on Research and Development in Information Retrieval, pp. 1227–1230. ACM, New York (2014)
2. Alayrac, J.B., Bojanowski, P., Agrawal, N., Sivic, J., Laptev, I., Lacoste-Julien, S.: Unsupervised learning from narrated instruction videos. In: Proceedings of the IEEE Conference on Computer Vision and Pattern Recognition, pp. 4575–4583. IEEE, Piscataway (2016)
3. Anne Hendricks, L., Wang, O., Shechtman, E., Sivic, J., Darrell, T., Russell, B.: Localizing moments in video with natural language. In: Proceedings of the IEEE Conference on Computer Vision and Pattern Recognition, pp. 5803–5812. IEEE, Piscataway (2017)
4. Bojanowski, P., Lajugie, R., Grave, E., Bach, F., Laptev, I., Ponce, J., Schmid, C.: Weakly-supervised alignment of video with text. In: Proceedings of the IEEE International Conference on Computer Vision, pp. 4462–4470. IEEE, Piscataway (2015)
5. Cao, D., He, X., Nie, L., Wei, X., Hu, X., Wu, S., Chua, T.S.: Cross-platform app recommendation by jointly modeling ratings and texts. ACM Trans. Inf. Syst. **35**(4), 37 (2017)
6. Cao, D., Nie, L., He, X., Wei, X., Zhu, S., Chua, T.S.: Embedding factorization models for jointly recommending items and user generated lists. In: Proceedings of the International ACM SIGIR Conference on Research and Development in Information Retrieval, pp. 585–594. ACM, New York (2017)
7. Feng, F., He, X., Liu, Y., Nie, L., Chua, T.S.: Learning on partial-order hypergraphs. In: Proceedings of the International Conference on World Wide Web, pp. 1523–1532. International World Wide Web Conferences Steering Committee (2018)
8. Frome, A., Corrado, G.S., Shlens, J., Bengio, S., Dean, J., Mikolov, T., et al.: Devise: a deep visual-semantic embedding model. In: Proceedings of the Advances in Neural Information Processing Systems, pp. 2121–2129. NIPS (2013)
9. Gaidon, A., Harchaoui, Z., Schmid, C.: Actom sequence models for efficient action detection. In: Proceedings of the IEEE Conference on Computer Vision and Pattern Recognition, pp. 3201–3208. IEEE, Piscataway (2011)
10. Gaidon, A., Harchaoui, Z., Schmid, C.: Temporal localization of actions with actoms. IEEE Trans. Pattern Anal. Mach. Intell. **35**(11), 2782–2795 (2013)
11. Gao, J., Sun, C., Yang, Z., Nevatia, R.: Tall: temporal activity localization via language query. In: Proceedings of the IEEE International Conference on Computer Vision, pp. 5267–5275. IEEE, Piscataway (2017)
12. Gao, J., Yang, Z., Sun, C., Chen, K., Nevatia, R.: Turn tap: temporal unit regression network for temporal action proposals. In: Proceedings of the IEEE International Conference on Computer Vision, pp. 3628–3636. IEEE, Piscataway (2017)
13. Hu, R., Xu, H., Rohrbach, M., Feng, J., Saenko, K., Darrell, T.: Natural language object retrieval. In: Proceedings of the IEEE Conference on Computer Vision and Pattern Recognition, pp. 4555–4564. IEEE, Piscataway (2016)
14. Karpathy, A., Fei-Fei, L.: Deep visual-semantic alignments for generating image descriptions. In: Proceedings of the IEEE Conference on Computer Vision and Pattern Recognition, pp. 3128–3137. IEEE, Piscataway (2015)

15. Kiros, R., Zhu, Y., Salakhutdinov, R.R., Zemel, R., Urtasun, R., Torralba, A., Fidler, S.: Skip-thought vectors. In: Proceedings of the Advances in Neural Information Processing Systems, pp. 3294–3302. NIPS (2015)
16. Lin, D., Fidler, S., Kong, C., Urtasun, R.: Visual semantic search: retrieving videos via complex textual queries. In: Proceedings of the IEEE Conference on Computer Vision and Pattern Recognition, pp. 2657–2664. IEEE, Piscataway (2014)
17. Pennington, J., Socher, R., Manning, C.: Glove: global vectors for word representation. In: Proceedings of the Conference on Empirical Methods in Natural Language Processing, pp. 1532–1543. ACL (2014)
18. Ren, S., He, K., Girshick, R., Sun, J.: Faster R-CNN: towards real-time object detection with region proposal networks. In: Proceedings of the Advances in Neural Information Processing Systems, pp. 91–99. NIPS (2015)
19. Rohrbach, M., Regneri, M., Andriluka, M., Amin, S., Pinkal, M., Schiele, B.: Script data for attribute-based recognition of composite activities. In: Proceedings of the European Conference on Computer Vision, pp. 144–157. Springer, Berlin (2012)
20. Sener, O., Zamir, A.R., Savarese, S., Saxena, A.: Unsupervised semantic parsing of video collections. In: Proceedings of the IEEE International Conference on Computer Vision, pp. 4480–4488. IEEE, Piscataway (2015)
21. Shou, Z., Wang, D., Chang, S.F.: Temporal action localization in untrimmed videos via multi-stage CNNs. In: Proceedings of the IEEE Conference on Computer Vision and Pattern Recognition, pp. 1049–1058. IEEE, Piscataway (2016)
22. Simonyan, K., Zisserman, A.: Very deep convolutional networks for large-scale image recognition (2014). arXiv:1409.1556
23. Singh, B., Marks, T.K., Jones, M., Tuzel, O., Shao, M.: A multi-stream bi-directional recurrent neural network for fine-grained action detection. In: Proceedings of the IEEE Conference on Computer Vision and Pattern Recognition, pp. 1961–1970. IEEE, Piscataway (2016)
24. Socher, R., Karpathy, A., Le, Q.V., Manning, C.D., Ng, A.Y.: Grounded compositional semantics for finding and describing images with sentences. Trans. Assoc. Comput. Linguist. **2**, 207–218 (2014)
25. Tran, D., Bourdev, L., Fergus, R., Torresani, L., Paluri, M.: Learning spatiotemporal features with 3d convolutional networks. In: Proceedings of the IEEE International Conference on Computer Vision, pp. 4489–4497. IEEE, Piscataway (2015)
26. Wang, X., He, X., Feng, F., Nie, L., Chua, T.S.: Tem: tree-enhanced embedding model for explainable recommendation. In: Proceedings of the International Conference on World Wide Web, pp. 1543–1552. International World Wide Web Conferences Steering Committee (2018)
27. Wang, X., Nie, L., Song, X., Zhang, D., Chua, T.S.: Unifying virtual and physical worlds: learning toward local and global consistency. ACM Trans. Inf. Syst. **36**(1), 4:1–4:26 (2017)

Chapter 3
Semantic Alignment Video Moment Localization

3.1 Introduction

In the traditional video retrieval task, the queries are often simple keywords that refer to desired actions, objects, or attributes. In contrast, video moment localization involves much more complex queries, ranging from simple phrases to complete sentences. As shown in Fig. 3.1, for example, the query "A Ferris wheel first comes into view" requires not only detecting the appearance of a "Ferris wheel" but also identifying the specific temporal relationship described by the word "first". A model that only localizes "Ferris wheel" is insufficient, as this object may appear multiple times in the video. Thus, effectively resolving this query requires not only locating the moment involving a "Ferris wheel" but also ensuring that it is the initial occurrence. The primary challenge of temporal moment localization, therefore, lies in comprehending complex query information and identifying the most relevant words for precisely localizing the desired moment.

Several studies [1, 6] have been proposed to address complex queries in video moment localization. These methods typically feed the entire description into a pre-trained language model (e.g., Skip-thoughts [14]) or a recurrent model (e.g., LSTM [9]) to generate a single feature vector representing the query. While this approach has shown some success, treating queries holistically can obscure critical keywords that carry important temporal and semantic cues. For instance, in Fig. 3.1, the word "first" is essential for identifying the correct temporal moment, distinguishing the target "Ferris wheel" from other similar instances. This indicates that the relationship between textual components and temporal moments has not been fully explored, highlighting the need for an adaptive language processing model that can selectively focus on key query components based on video context.

In this work, we aim to address this gap by incorporating a language processing module that captures the spatial-temporal information essential for moment localization. Specifically, we focus on exploiting the correlation between textual and

M. Liu et al., *Intelligent and Efficient Video Moment Localization*,
https://doi.org/10.1007/978-3-031-87588-5_3

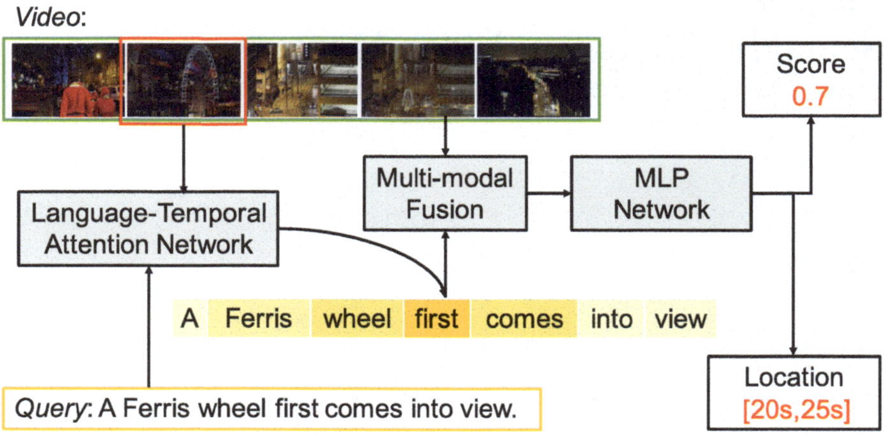

Fig. 3.1 The pipeline of our proposed model

visual features, enabling the model to highlight the critical components of the query that are most useful for localizing the desired moment.

To this end, we propose the cross-modal moment localization network (ROLE), which jointly learns the representation of the query and performs video moment localization, as shown in Fig. 3.1. First, we introduce a language-temporal attention module that generates effective query representations by adaptively re-weighting each word's importance based on both textual and temporal context. This allows the model to focus on the most relevant parts of the query, improving robustness against irrelevant variations in the query. Next, we employ a multi-modal processing module to jointly model the query and temporal context features. Finally, we train a multi-layer perceptron (MLP) network to estimate the relevance score and identify the temporal boundaries of the target moment.

Extensive experiments conducted on two public datasets, i.e., DiDeMo and Charades-STA, demonstrate that our model significantly outperforms state-of-the-art baselines. Additionally, we release our code to the research community for further exploration.[1]

3.2 Related Work

This work is related to grounding referential expressions, temporal action localization, and language grounding in video.

[1] https://acmmm18.wixsite.com/role.

Grounding Referential Expressions Grounding referential expressions [18, 21, 39, 41] refers to the task of localizing an image region described by a given referring expression, which is typically formulated as a retrieval problem over image regions. The process usually begins by segmenting each image into region proposals [2, 15, 34, 42], followed by scoring each proposal relative to the query expression. The proposal with the highest score is selected as the grounded result.

Considerable efforts have been made to improve the computation of the matching score between each proposal and the referring expression. Mao et al. [20] proposed a model that jointly considers local candidate features and the overall image context to predict the matching score for each proposal. However, relying solely on these features often proves insufficient for accurate matching. To address this, Yu et al. [38] incorporated contextual features from other regions within the image, significantly improving performance by integrating these additional features into the model. However, this approach fails to learn explicit correspondences between the components of the referring expression and the entities in the image. Recent studies have parsed language expressions into distinct components and aligned these components with image regions in an end-to-end fashion. For example, Hu et al. [12] parsed referring expressions into subject, relation, and object components, aligning these components with image regions using a modular neural architecture. Similarly, Yu et al. [40] decomposed expressions into components related to the subject appearance, location, and relationships to other objects, using a soft attention mechanism to facilitate this alignment.

While these models have been effective in grounding referential expressions in images, extending them to video moment localization introduces significant challenges. These models struggle with capturing temporal information, which is a key aspect distinguishing video analysis from static image analysis.

Temporal Action Localization Temporal action localization focuses on predicting the start and end times of a specific action in long, untrimmed videos [5, 17, 19]. Early work, such as that by Sun et al. [31], addressed fine-grained action localization using untrimmed web videos by transferring image labels into their model. Shou et al. [26] extended this work by using multi-stage 3D convolutional networks (ConvNets) for temporal action localization in complex, untrimmed videos. Ma et al. [19] proposed ranking losses within the RNN learning objective to better capture action progression, while Singh et al. [29] extended the two-stream framework [27] by incorporating features from bounding boxes around actors, in addition to full-frame features. More recently, Gao et al. [7] introduced a temporal unit regression network that predicts action proposals and refines temporal boundaries using temporal coordinate regression.

Despite these advances, many temporal action localization methods are restricted to predefined action categories, limiting their flexibility in dealing with a broader range of activities. To address this, Gao et al. [6] proposed a cross-modal temporal regression localizer that jointly models natural language queries and video moments, extending the capabilities of action localization. Similarly, Hendricks et al. [1] developed a moment context network that uses both local and global video features to localize language queries in video. Although these models show strong performance, they typically encode the entire query as a single feature vector, which may not fully capture the intricate information conveyed in complex queries.

Language Grounding in Video Unlike video retrieval [3, 10, 13, 36], which involves retrieving a video from a set of candidates based on a natural language query [22, 33, 35], language grounding in video focuses on aligning textual phrases with temporal segments of the video [8, 25, 30]. The body of work in this area is still developing, and existing studies often face challenges related to a limited natural language vocabulary.

Tellex et al. [32] proposed a model to retrieve video segments from home surveillance footage using queries containing a fixed set of spatial prepositions. Yu et al. [37] worked with a limited vocabulary consisting of four objects and four verbs, learning word representations from short video clips paired with sentences. Lin et al. [16] presented a model that locates objects in video by parsing descriptions into semantic graphs and matching these graphs to visual concepts using linear programming. Regneri et al. [24] introduced a corpus that aligns high-quality videos with multiple natural language descriptions of exhibited actions. Bojanowski et al. [4] proposed a method to automatically provide timestamps for each sentence, effectively aligning video content with natural language descriptions.

However, in contrast to the typical video-text alignment tasks where a video is paired with multiple sentences in a temporal sequence, our task focuses on localizing a single query in a video. Many previous methods, which are designed for aligning instructions with videos, are restricted to structured video content where temporal ordering of instructions is guaranteed. This approach is not easily applicable to general, unstructured video content, where the goal is to localize a specific moment in response to a single, often complex query.

3.3 Methodology

In this section, we first formulate the task of video moment localization. Then, we introduce two state-of-the-art models that serve as fundamental components of our approach. Finally, we present our proposed language-temporal attention network, designed to enhance the localization process by focusing on effective word-level attention based on the temporal context of the video.

3.3.1 Problem Formulation

The task of video moment localization involves identifying a specific temporal segment (i.e., moment) within a video, given a natural language description [1, 6].

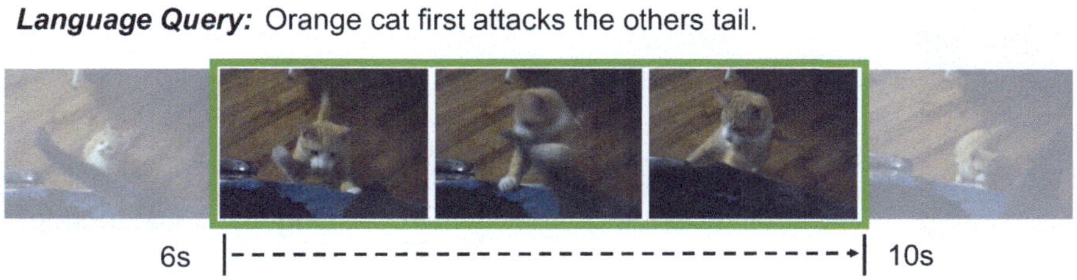

Fig. 3.2 An illustration of video moment localization

These moments are typically characterized by start and end time points that define the segment of interest in the video. More formally, the objective is to find the start time τ_s and end time τ_e of the moment that best matches the language query Q.

Given a video $\mathcal{V} = \{f_1, f_2, \ldots, f_N\}$, where each f_i represents the i-th frame, and a language query Q, the task is to predict the time window $[\tau_s, \tau_e]$ that corresponds to the most relevant segment of the video. The video is usually divided into candidate moments $C = \{c_1, c_2, \ldots, c_M\}$, where each moment c_i is assigned a temporal bounding box $[t_s, t_e]$ using a sliding window strategy [1, 6]. The goal is to estimate the relevance score between each candidate moment and the query Q.

To illustrate this process, consider the example in Fig. 3.2, where the video depicts an orange cat interacting with a black cat. The query provided is, "orange cat first attacks the other's tail," and the objective is to localize the moment where the orange cat first attacks the black cat's tail. The expected output would be the predicted start time (6 seconds) and end time (10 seconds), represented by the green bounding box.

3.3.2 *Preliminaries*

To effectively match the query and the candidate moments, a straightforward approach is to map the visual features of the candidate moments and the textual features of the query into a common embedding space and minimize the distance for positive moment-query pairs. Motivated by this intuition, two state-of-the-art methods have been proposed: the Moment Context Network (MCN) [1] and the Cross-modal Temporal Regression Localizer (CTRL) [6], both of which are shown in Fig. 3.3.

The MCN is a joint video-language model that encourages features of the query and video to be close within a shared embedding space. Similarly, the CTRL is

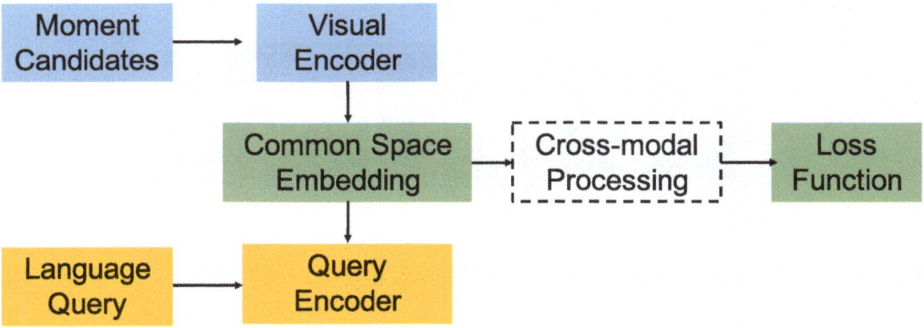

Fig. 3.3 The unified framework of the existing video moment localization methods. The dashed line indicates the unique module of the CTRL model

a cross-modal localization model that jointly models textual queries and video moments. These two models, however, differ in several key aspects:

1. **Context Information**: Both models concatenate the visual features of the current moment with its contextual moments to form a single feature vector, which is then linearly transformed to create the final visual representation. MCN takes into account the entire set of candidate moments as context, while CTRL includes only the immediate preceding and succeeding moments as context.
2. **Query Encoder**: MCN uses the final output of an LSTM network to represent the query, whereas CTRL employs an offline Skip-Thoughts feature [14] to encode the query.
3. **Cross-Modal Processing**: CTRL uses a multi-modal fusion method to integrate the query and moment features within the common embedding space, whereas MCN directly calculates the distance between query and moment features in the shared space.

While these models perform well compared to the baselines, they treat the query as a holistic unit, thereby overlooking the distinct contributions of specific keywords that convey important spatial-temporal cues. Depending on the particular target moment, different words in the query may play a more significant role in the estimation process. For instance, if the desired moment is "a girl in a red coat" among many people, words like "girl" and "red" are critical for accurate localization. If the girl appears multiple times in the video, and the query becomes "the girl in red first appears," then the contextual word "first" should play a crucial role in locating the target moment.

Thus, integrating an attention mechanism into the moment retrieval model is a natural approach to dynamically attend to each word in the query, depending on the temporal context information. This enables the model to focus on the words most relevant for accurately localizing the desired moment. Specifically, given a moment candidate c_i and a language query Q, we first employ an attention network to adaptively reweight the importance of each word in the query based on the moment's context. Subsequently, we use a multi-modal processing module to fuse the representations of the query and the candidate moment. Finally, a multi-layer perceptron (MLP) is applied to compute the relevance score, which measures the compatibility between c_i and Q for temporal localization of the moment.

3.3.3 Language-Temporal Attention Network

Our attention network, as shown in Fig. 3.4, is designed to effectively integrate word-level information from the query with temporal context from the video. For a given query Q composed of a sequence of T words $\{w_t\}_{t=1}^{T}$, each word w_t is first projected into an embedding vector $\mathbf{e}_t$ using Glove embeddings [23]. A Bi-directional LSTM (Bi-LSTM) is then employed to encode the entire query, with the sequence of word embeddings $\{\mathbf{e}_t\}_{t=1}^{T}$ as input. The Bi-LSTM outputs a forward hidden state $\mathbf{h}_t^{fw}$ and a backward hidden state $\mathbf{h}_t^{bw}$ at each time t. These are concatenated to form the combined hidden state $\mathbf{h}_t$, which captures information from the word w_t along with its preceding and succeeding context words. The formulation is as follows:

$$\begin{cases} \mathbf{e}_t = embedding(w_t), \\ \mathbf{h}_t^{fw} = LSTM^{fw}(\mathbf{e}_t, \mathbf{h}_{t-1}^{fw}), \\ \mathbf{h}_t^{bw} = LSTM^{bw}(\mathbf{e}_t, \mathbf{h}_{t-1}^{bw}), \\ \mathbf{h}_t = [\mathbf{h}_t^{fw}, \mathbf{h}_t^{bw}]. \end{cases} \tag{3.1}$$

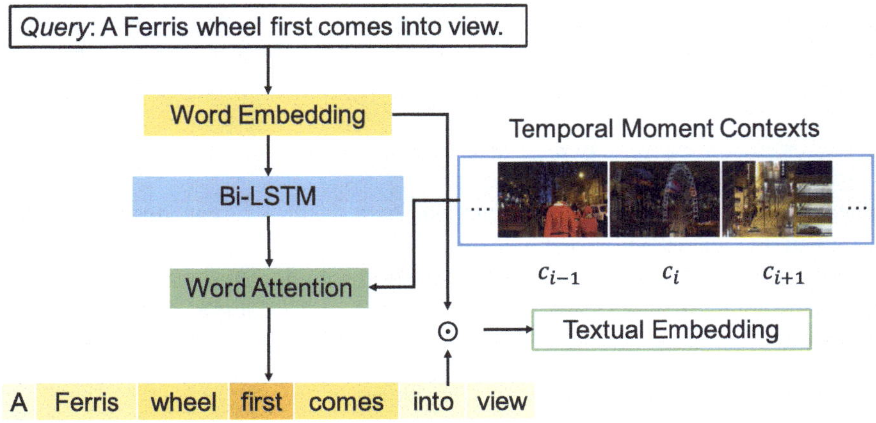

Fig. 3.4 Illustration of the language-temporal attention network

Although this formulation effectively captures the context surrounding each word, it assumes that all words contribute equally to the query embedding, ignoring the scenario-specific importance of individual words. As previously mentioned, such a uniform treatment of words may fail to distinguish between the desired moment and moments with similar visual features.

To address this issue, we integrate the temporal moment context[2] into our attention mechanism, enabling the model to assign higher importance scores to words that are more useful for localizing the target moment. Specifically, given the hidden states $\mathcal{H} = \{\mathbf{h}_t\}_{t=1}^{T}$, input moment c_i, and its temporal moment context c_j (where $j \in \{i-n, \ldots, i-1, i+1, \ldots, i+n\}$ and n represents the neighbor size), we define the attention model as follows:

$$\begin{cases} \mathbf{r}_t = f(\mathbf{W}_q \mathbf{h}_t + \sum_{j=i-n}^{i+n} \mathbf{W}_c \mathbf{x}_{c_j} + \mathbf{b}), \\ a_t = \frac{\boldsymbol{\beta}^T \mathbf{r}_t}{\sum_{t=1}^{T} \boldsymbol{\beta}^T \mathbf{r}_t}, \end{cases} \tag{3.2}$$

where $\mathbf{x}_{c_j}$ is the feature vector for each moment candidate, $\boldsymbol{\beta}$ is a trainable vector, $\mathbf{W}_q$ and $\mathbf{W}_c$ are embedding matrices for the textual and temporal components respectively, $\mathbf{b}$ is a bias vector, and f is the Rectified Linear Unit (ReLU, $f(x) = \max(0, x)$) function.

Once the attention scores for each word in the query are established, the query representation can be constructed as:

$$\mathbf{q} = \sum_{t=1}^{T} a_t \mathbf{e}_t. \tag{3.3}$$

Having derived the textual embedding $\mathbf{q}$ as well as the temporal embeddings of the current input moment and its neighboring moments $\mathbf{x}_{c_j}$, we can obtain a cross-modal representation for the current query-moment pair by using the concatenation operator:

$$\mathbf{x}_{c,q} = \mathbf{x}_{c_{i-n}} \oplus \ldots \oplus \mathbf{x}_{c_{i-1}} \oplus \mathbf{x}_{c_i} \oplus \mathbf{x}_{c_{i+1}} \oplus \ldots \oplus \mathbf{x}_{c_{i+n}} \oplus \mathbf{q}, \tag{3.4}$$

where $\oplus$ denotes the vector concatenation. As a result, $\mathbf{x}_{c,q}$ encodes information from both the textual and visual modalities.

Next, we pass $\mathbf{x}_{c,q}$ through a MLP. The MLP, with its non-linear hidden layers, facilitates complex interactions among the cross-modal features, which enhances the expressiveness of the representation. The prediction layer produces the relevance

[2] In this paper, temporal moment context refers to the n-neighbor moment candidates surrounding the current moment.

score s_{cq} for the moment-query pair (c, q), as well as the location offset $[t_s - \tau_s, t_e - \tau_e]$ between the current moment and the ground truth:

$$\begin{cases} \mathbf{o}_1 = f_1(\mathbf{W}_1\mathbf{x}_{c,q} + \mathbf{b}_1), \\ \mathbf{o}_2 = f_2(\mathbf{W}_2\mathbf{o}_1 + \mathbf{b}_2), \\ \quad\vdots \\ \mathbf{o}_L = f_L(\mathbf{W}_L\mathbf{o}_{L-1} + \mathbf{b}_L), \end{cases} \tag{3.5}$$

where $\mathbf{W}_l$, $\mathbf{b}_l$, f_l, and $\mathbf{o}_l$ denote the weight matrix, bias vector, activation function, and the output vector of the l-th hidden layer, respectively. ReLU units are used as activation functions for each hidden layer. The final output vector $\mathbf{o}_L = [s_{cq}, \delta_s, \delta_e]$ consists of the matching score s_{cq} as well as the location offsets $\delta_s = t_s - \tau_s$ and $\delta_e = t_e - \tau_e$.

3.3.4 Learning

The overall optimization objective is defined as:

$$L = L_{align} + \lambda L_{location}, \tag{3.6}$$

where λ is a hyperparameter that balances the contributions of the alignment and localization regression losses.

Alignment Loss Similar to [6], we formulate the alignment task as a binary classification problem. Given a set of moment candidates C extracted from a video $\mathcal{V}$ and a query Q, we divide the moment-query pairs into two groups: the aligned pairs $\mathcal{P}$ and the misaligned pairs $\mathcal{N}$. The alignment loss encourages the aligned pairs to have positive scores and the misaligned pairs to have negative scores:

$$L_{align} = \sum_{(c,q)\in\mathcal{P}} \lambda_1 \log(1 + \exp(-s_{cq})) + \sum_{(c,q)\in\mathcal{N}} \lambda_2 \log(1 + \exp(s_{cq})), \tag{3.7}$$

where λ_1 and λ_2 are two hyperparameters that balance the weights between positive and negative pairs.

Location Loss Since the bounding box $[t_s, t_e]$ of the positive moment candidates may not exactly match the ground truth $[\tau_s, \tau_e]$, we also consider the location offset between the positive candidates and the ground truth. Let $[\delta_s^*, \delta_e^*]$ represent the ground truth offsets. The location offset regression is formulated as follows:

$$L_{location} = \sum_{(c,q)\in\mathcal{P}} |\delta_s - \delta_s^*| + |\delta_e - \delta_e^*|. \tag{3.8}$$

During the training phase, the offset regression loss is only applied to positive samples. In the testing phase, once we identify the moment candidate with the highest alignment score, we apply the predicted offset values to adjust the predicted temporal boundaries, thereby bringing the final temporal bounding box closer to the ground truth.

3.4 Experiments

We first evaluated the effectiveness of our proposed model on two video moment localization datasets: DiDeMo and Charades-STA. We then investigated the impact of our well-designed attention network on localization performance.

3.4.1 Experimental Settings

3.4.1.1 Datasets

We evaluated our proposed ROLE model on two benchmark datasets: DiDeMo and Charades-STA.

DiDeMo [1] This dataset contains distinct video moments paired with natural language descriptions to uniquely localize moments within videos. It includes over 10,000 personal videos, each lasting between 25 to 30 seconds, and features more than 40,000 localized text descriptions. The videos in DiDeMo are collected from Flickr, representing a diverse set of real-world content, including both interesting and uneventful moments, which may be excluded from edited videos. Each video in the released dataset is divided into six five-second segments, and each segment is represented by a 4096-dimensional VGG [28] feature vector. For language features, the dataset uses 300-dimensional GloVe [23] embeddings to represent each word.

Charades-STA [6] This dataset is based on the Charades dataset and contains approximately 10,000 videos, each annotated with temporal activity and multiple video-level descriptions. To facilitate temporal activity localization, a semi-automated approach was proposed in [6] to generate clip-level sentence annotations, resulting in the Charades-STA dataset. Charades-STA consists of 6672 videos. Since the released Charades-STA dataset only includes video-description files, we downloaded the videos from the provided website[3] and extracted features for each moment candidate.[4] Specifically, each video was segmented into temporal units with a window size of 16 frames and an overlap of 12 frames. We then extracted

[3] http://allenai.org/plato/charades/.

[4] For our experiments, we used videos scaled to 480p as the input.

Table 3.1 Statistics of the Charades-STA and DiDeMo datasets

Dataset	# Videos	# Queries	Domain	Video Source
Charades-STA	6, 672	16, 128	Homes	Daily Activities
DiDeMo	10, 464	40, 543	Open	Flickr

C3D features[5] for each temporal unit and constructed moment candidates with different unit sizes [4, 8, 16, 32]. The temporal feature of each moment candidate was obtained by mean pooling the features of the corresponding units.

The statistics for both datasets are summarized in Table 3.1. The reported experimental results in this paper are based on the aforementioned datasets.[6] Furthermore, we conducted our experiments using TensorFlow, and selected AdamOptimizer as the optimization algorithm. The model was trained on a server equipped with 16 Tesla K80 GPUs.

3.4.1.2 Evaluation Metrics

To comprehensively evaluate our model and the baselines, we adopted the "R@n, IoU=m" metric as proposed by Hu et al. [11]. Specifically, given a language query, this metric calculates the percentage of the top-n results whose Intersection over Union (IoU) is greater than m. In the following discussion, we used $R(n, m)$ to denote "R@n, IoU=m".

This metric is applied at the query level, and the overall performance is reported as the average across all queries:

$$R(n, m) = \frac{1}{N_q} \sum_{i=1}^{N_q} r(n, m, q_i), \tag{3.9}$$

where $r(n, m, q_i)$ represents the recall for a query q_i, N_q is the total number of queries, and $R(n, m)$ is the averaged performance over all queries.

3.4.1.3 Baselines

We compared our proposed method with the following state-of-the-art baselines:

- **MCN** [1]: This model is designed to localize natural language queries within videos by integrating both local and global video features. The query features are

---

[5] https://github.com/facebook/C3D.

[6] In the experiments, we set the context moment number n to 1. The context window length was set to 128 frames for the Charades-STA dataset and 5 seconds for the DiDeMo dataset.

extracted using an LSTM model. MCN operates on the assumption that features from the given query and corresponding video moments should be similar in a shared embedding space, with a loss function enforcing similarity between these features. The details of this model are provided in Sect. 3.3.2.

- **CTRL** [6]: This is a cross-modal temporal localization model developed to localize activities in videos based on natural language queries. It jointly models the textual query using Skip-Thoughts and video moments, producing a similarity score along with predicted action boundaries. Further details on CTRL are provided in Sect. 3.3.2.

3.4.2 Performance Comparison

We conducted an empirical study to evaluate whether our proposed model can achieve improved localization performance compared to existing methods. The results on the two datasets, Charades-STA and DiDeMo, are presented in Tables 3.2 and 3.3, respectively. Several observations can be drawn from these results:

- MCN performs worse than the other baselines, likely due to its approach of treating the entire set of moments as the context for each candidate moment, which can introduce noisy features and lead to negative transfer. Additionally, since MCN models the relationship between the query and moment features by simply minimizing their distance in a shared embedding space, the cross-modal relationships are not fully exploited. Furthermore, the use of an LSTM network

Table 3.2 Performance comparison between our proposed model and the state-of-the-art baselines on the Charades-STA (p-value*: p-value over $R(1, 0.5)$)

Method	R@1 IoU=0.5	R@1 IoU=0.3	R@1 IoU=0.1	R@5 IoU=0.5	R@5 IoU=0.3	R@5 IoU=0.1	p-value*
MCN	4.05%	13.57%	31.26%	19.67%	50.53%	77.60%	5.30E-09
CTRL	9.30%	21.45%	35.05%	33.41%	65.59%	80.43%	1.80E-02
ROLE	**12.12%**	**25.26%**	**37.39%**	**40.59%**	**70.13%**	**82.82%**	–

The best experimental results are highlighted in bold

Table 3.3 Performance comparison between our proposed model and the state-of-the-art baselines on the DiDeMo (p-value*: p-value over $R(1, 0.5)$)

Method	R@1 IoU=0.5	R@1 IoU=0.7	R@1 IoU=0.9	R@5 IoU=0.5	R@5 IoU=0.7	R@5 IoU=0.9	p-value*
MCN	23.32%	15.35%	15.31%	41.03%	20.37%	19.77%	3.68E-09
CTRL	26.45%	15.45%	15.38%	68.11%	29.00%	26.12%	5.52E-04
ROLE	**29.40%**	**15.68%**	**15.55%**	**70.72%**	**33.08%**	**29.73%**	–

The best experimental results are highlighted in bold

to encode the query results in an inability to identify distinctive words, as it treats all words uniformly.
- CTRL achieves better localization results compared to MCN. This improvement is due to its incorporation of neighboring moments as contextual information, as well as its cross-modal processing component, which allows for more effective interactions between the visual and textual modalities. However, its performance is still limited by encoding the entire query as a single holistic feature. As discussed in Sect. 3.1, different words contribute differently to the localization process, and such coarse-grained encoding may obscure the varying importance of individual words.
- Our proposed model ROLE achieves the best performance, significantly surpassing all baselines. It consistently shows improvements over MCN and CTRL, which validates the importance of integrating temporal context into the query embedding and dynamically selecting distinctive words within the query. This approach enables a more nuanced understanding of the query and enhances localization accuracy.

It is worth noting that, for the DiDeMo dataset, since the positive moment-query pairs are well-aligned (i.e., there is no location offset between them), we only used the alignment loss for training both CTRL and ROLE.

In addition, we conducted significance tests comparing our model to each of the baselines using the $R(1, 0.5)$ results. The results indicate that all p-values are much smaller than 0.05, demonstrating that the performance advantage of our model over the baselines is statistically significant.

3.4.3 Ablation Study

We conducted an ablation study to further investigate the effectiveness of the language-temporal attention network by evaluating different variants of our model:

- **ROLE-NT**: In this variant, we removed the temporal context information component from the language-temporal attention model. Specifically, each word's attention value is determined solely by the query and the current moment, without taking into account neighboring context.
- **ROLE-NV**: In this variant, we replaced the language-temporal attention mechanism with a query attention model that only depends on the embeddings of the query words. In other words, all temporal visual information was excluded.

We compared these model variants on the Charades-STA and DiDeMo datasets. Figure 3.5 presents the results of the component-wise comparison.

- Analyzing the performance of ROLE-NV in Fig. 3.5a, b, we observed that removing the language-temporal attention significantly reduces the model's expressiveness, leading to degraded localization results, especially in terms of

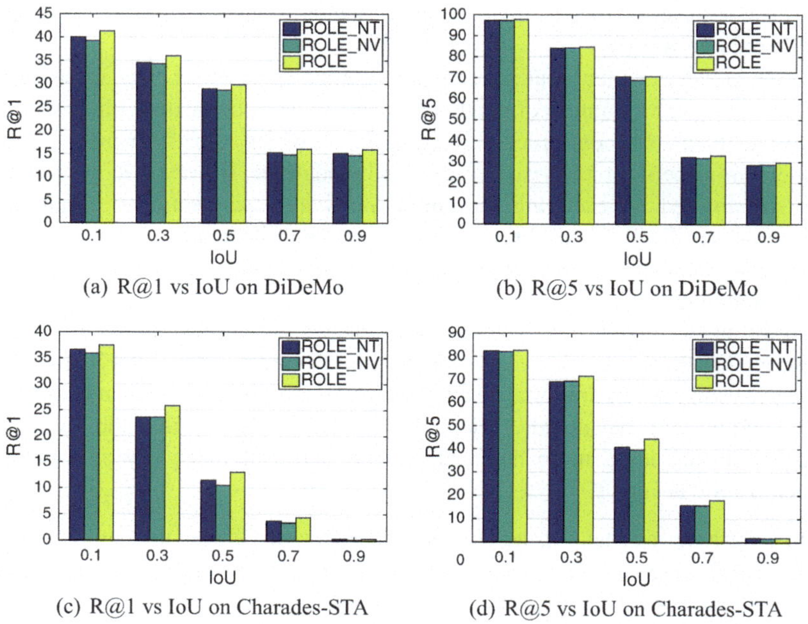

Fig. 3.5 Performance comparison among the variants of our proposed model over the DiDeMo and the Charades-STA datasets. From left to right: (**a**) The R@1 vs IoU $\in \{0.1, 0.3, 0.5, 0.7, 0.9\}$ on the DiDeMo dataset; (**b**) The R@5 vs IoU $\in \{0.1, 0.3, 0.5, 0.7, 0.9\}$ on the DiDeMo dataset; (**c**) the R@1 vs IoU $\in \{0.1, 0.3, 0.5, 0.7, 0.9\}$ on the Charades-STA dataset; and (**d**) the R@5 vs IoU $\in \{0.1, 0.3, 0.5, 0.7, 0.9\}$ on the Charades-STA dataset

R@1. This suggests that relying solely on the query information is insufficient to identify keywords related to the desired moments and to highlight cross-modal correlations effectively.

- ROLE-NT outperforms ROLE-NV, indicating that incorporating visual information from the current moment strengthens the comprehension of the given query. By leveraging cross-modal correlations, ROLE-NT enhances the overall expressiveness of the model.
- Our proposed model ROLE consistently outperforms both ROLE-NT and ROLE-NV, demonstrating the critical importance of incorporating visual features from the surrounding moment context. Removing visual features altogether (i.e., ROLE-NV) or only considering the visual features of the current moment (i.e., ROLE-NT) limits the model's ability to capture temporal correlations, thereby reducing the effectiveness of query-moment comprehension and degrading localization performance.

3.4.4 Attention Visualization

Beyond achieving superior performance, one of the key advantages of ROLE over other methods is its ability to distinguish the most relevant words for identifying the ground truth moment through language-temporal attention. To illustrate this capability, we presented several examples, visualizing their attention scores and moment localization results, as shown in Fig. 3.6.

Figure 3.6a depicts a scenario in which several people, including both women and men, are enjoying the scenery on top of a mountain–some sitting on stones, others standing on distant highlands. Given the query "Woman in red comes into view", we expect the retrieved moment to feature a woman wearing a red coat. The distinctive information needed for localizing the target moment is intuitively "woman" and "red". We fed the video and the query into ROLE, and consequently obtained the attention scores for each word in the query. Several noteworthy observations emerge:

- The word "woman" is marked with the darkest shade of orange, indicating that it attracts the most attention.
- The words "in" and "red" are marked in a slightly lighter shade of orange, reflecting moderate attention compared to "woman".
- The remaining words have the lowest attention values.

These findings align with our expectations and demonstrate that ROLE effectively adapts to identify the words that are most useful for localization, depending on the temporal moment context. This serves as evidence for the effectiveness of our language-temporal attention network.

Figure 3.6b provides another example where the video depicts a scene involving a man typing on a notebook, taking a drink from a glass of water, and then returning to typing. The query is "A person takes a drink from a glass of water", and we expect our model to accurately localize the action of "drinking water" among the activities in the video. From the attention scores visualized in Fig. 3.6b, we observed that the words related to the action "drinking water" receive more attention compared to other words. This is consistent with our previous analysis. Although the initial alignment result (highlighted by the green dashed box) is not perfect, it does capture the corresponding action moment. Furthermore, our model refines the alignment through the regression module, providing a more accurate result (shown by the blue dashed box). This justifies the effectiveness of our proposed attention mechanism and the regression loss.

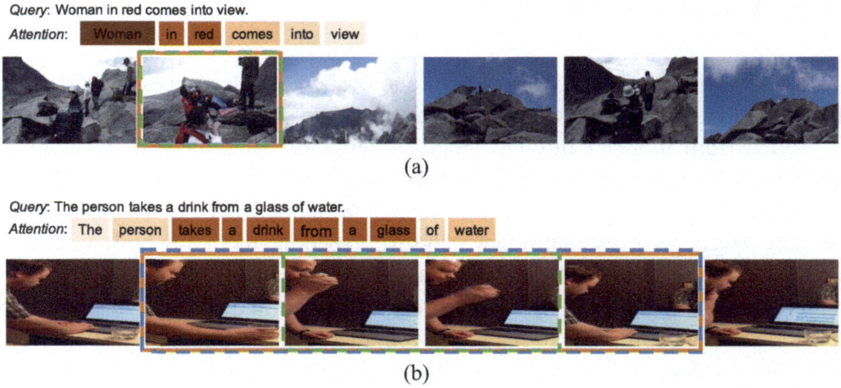

Fig. 3.6 Visualization of the language-temporal attention on DiDeMo and Charades-STA. The Ground truth moments are outlined in the orange box, while the green dash box shows our prediction with the alignment score and the blue dash box shows the localization result with our regression correction. The word attention is presented with different colors, and the darker color states the higher value. (**a**) The moment localization result on the DiDeMo. (**b**) The moment localization result on the Charades-STA

3.4.5 Qualitative Results

To gain deeper insights into our proposed ROLE model, we illustrated several moment localization results using different language queries. Specifically, examples from the DiDeMo and Charades-STA datasets are shown in Figs. 3.7 and 3.8, respectively. Additionally, we compared the localization results produced by the baselines.

Figure 3.7 depicts a scene involving a moving camera, where a small white car leaves the frame while another car remains stationary. We used the models to localize the moment corresponding to the query "The small white car is leaving the frame". Comparing the retrieval results from all the methods, we made the following observations:

- MCN simply returns a moment containing the "white car" from the set of candidate moments, without capturing the key action "leaving the frame". This could be due to two main reasons: (1) MCN uses the entire candidate set as a global feature to enhance the representation of the current moment. When most of the moments in the video relate to another car, the global feature fails to precisely represent the desired scene. (2) The model adopts the last state of an LSTM as the textual embedding for the query, which is insufficient for identifying crucial keywords like "leaving".
- As shown in Fig. 3.7c, CTRL provides an unsatisfactory result by returning a moment that includes both cars. Compared to MCN, CTRL improves localization by integrating neighboring moments as context rather than using the entire background. However, it captures only part of the desired moment, since the

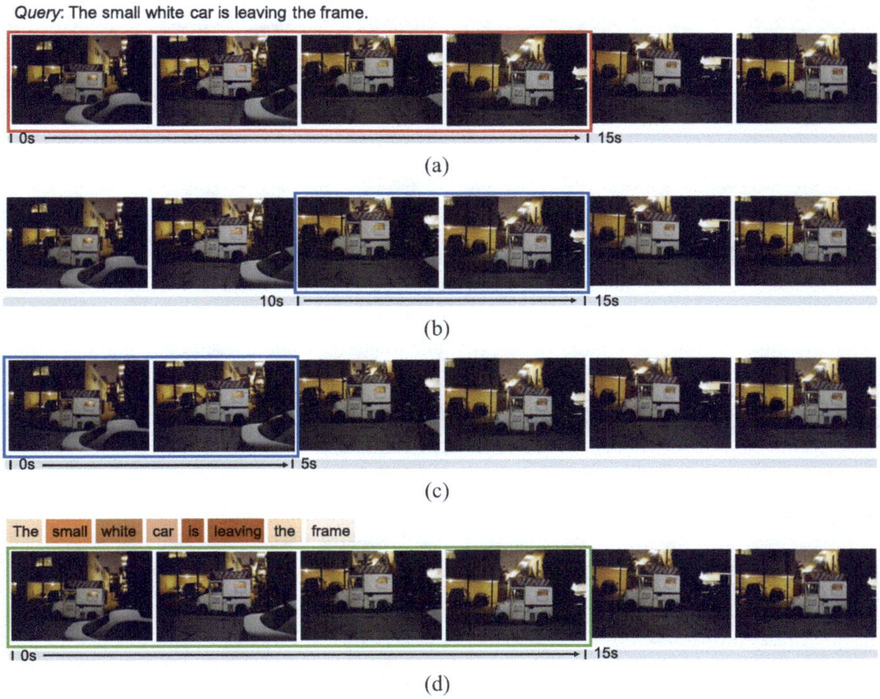

Fig. 3.7 Moment localization results on the DiDeMo dataset. All the above figures are the R@1 results. The gray bar denotes the timeline. The red, blue, and green bounding boxes denote the ground truth, the result of the baseline model, and our proposed model, respectively. (**a**) The desired moment of the moment localization. (**b**) The moment localization result of the MCN. (**c**) The moment localization result of the CTRL. (**d**) The moment localization result of the ROLE

query is encoded using the offline Skip-Thoughts tool, which overlooks the sequential relation "leaving". As a result, CTRL produces a suboptimal moment.

- Our proposed model ROLE outperforms the baselines, as depicted in Fig. 3.7d. The localized moment, with the query attention, demonstrates that ROLE can capture both the desired entity "small white car" and the sequential action "leaving". This result again verifies the effectiveness of our proposed video moment localization model.

Similarly, in the example shown in Fig. 3.8, our model produces more accurate results compared to the other models. In this video, there are two individuals, i.e., one is getting dressed while the other is engaged in another activity. The word "dressing" is the most critical information for distinguishing the desired moment, as we expected. Since MCN and CTRL model the language query holistically, they treat all words as having equal importance, which can introduce noise into the textual representation. Additionally, these models struggle to identify distinctive words like "dressing" from the query.

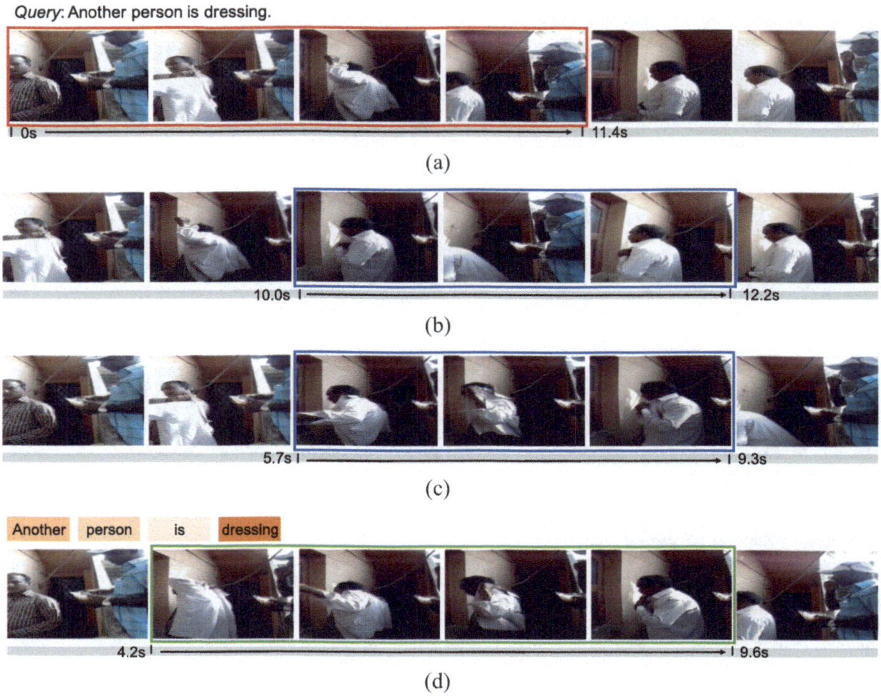

Fig. 3.8 Moment localization results on the Charades-STA dataset. All the above figures are the R@1 results. The gray bar denotes the timeline. The red, blue, and green bounding boxes denote the ground truth, the result of the baseline model, and our proposed model, respectively. (**a**) The desired moment of the moment localization. (**b**) The moment localization result of the MCN. (**c**) The moment localiation result of the CTRL. (**d**) The moment localization result of the ROLE

As a result, MCN returns a moment in which the person is simply finishing getting dressed, while CTRL selects a moment that includes only the latter part of the dressing action. In contrast, the moment localized by our proposed model has the highest IoU with the ground truth, indicating a more accurate localization. However, it does not perfectly match the ground truth. This discrepancy may be due to the coarse-grained sliding window approach we used to generate moment candidates for the Charades-STA dataset.

3.5 Summary

In this chapter, we introduce a cross-modal moment localization model designed to localize specific moments in a video based on a natural language description. To effectively capture the explicit correspondence between textual components and temporal moments, we propose a language-temporal attention network that adaptively identifies the most relevant words based on the temporal context of the video. We conduct extensive experiments on two publicly available benchmark datasets, demonstrating the superior performance and effectiveness of our proposed model.

References

1. Anne Hendricks, L., Wang, O., Shechtman, E., Sivic, J., Darrell, T., Russell, B.: Localizing moments in video with natural language. In: Proceedings of the IEEE Conference on Computer Vision and Pattern Recognition, pp. 5803–5812. IEEE, Piscataway (2017)
2. Arbeláez, P., Pont-Tuset, J., Barron, J.T., Marques, F., Malik, J.: Multiscale combinatorial grouping. In: Proceedings of the IEEE Conference on Computer Vision and Pattern Recognition, pp. 328–335. IEEE, Piscataway (2014)
3. Aslandogan, Y.A., Yu, C.T.: Techniques and systems for image and video retrieval. IEEE Trans. Knowl. Data Eng. **11**(1), 56–63 (1999)
4. Bojanowski, P., Lajugie, R., Grave, E., Bach, F., Laptev, I., Ponce, J., Schmid, C.: Weakly-supervised alignment of video with text. In: Proceedings of the IEEE International Conference on Computer Vision, pp. 4462–4470. IEEE, Piscataway (2015)
5. Escorcia, V., Heilbron, F.C., Niebles, J.C., Ghanem, B.: Daps: deep action proposals for action understanding. In: Proceedings of the European Conference on Computer Vision, pp. 768–784. Springer, Berlin (2016)
6. Gao, J., Sun, C., Yang, Z., Nevatia, R.: Tall: temporal activity localization via language query. In: Proceedings of the IEEE International Conference on Computer Vision, pp. 5267–5275. IEEE, Piscataway (2017)
7. Gao, J., Yang, Z., Chen, K., Sun, C., Nevatia, R.: Turn tap: temporal unit regression network for temporal action proposals. In: Proceedings of the IEEE International Conference on Computer Vision, pp. 3628–3636. IEEE, Piscataway (2017)
8. Guo, H., Wang, J., Xu, M., Zha, Z.J., Lu, H.: Learning multi-view deep features for small object retrieval in surveillance scenarios. In: Proceedings of the ACM International Conference on Multimedia, pp. 859–862. ACM, New York (2015)
9. Guo, Z., Gao, L., Song, J., Xu, X., Shao, J., Shen, H.T.: Attention-based LSTM with semantic consistency for videos captioning. In: Proceedings of the ACM International Conference on Multimedia, pp. 357–361. ACM, New York (2016)
10. Hu, W., Xie, D., Fu, Z., Zeng, W., Maybank, S.: Semantic-based surveillance video retrieval. IEEE Trans. Image Process. **16**(4), 1168–1181 (2007)
11. Hu, R., Xu, H., Rohrbach, M., Feng, J., Saenko, K., Darrell, T.: Natural language object retrieval. In: Proceedings of the IEEE Conference on Computer Vision and Pattern Recognition, pp. 4555–4564. IEEE, Piscataway (2016)
12. Hu, R., Rohrbach, M., Andreas, J., Darrell, T., Saenko, K.: Modeling relationships in referential expressions with compositional modular networks. In: Proceedings of the IEEE Conference on Computer Vision and Pattern Recognition, pp. 4418–4427. IEEE, Piscataway (2017)
13. Jiang, Y.G., Ngo, C.W., Yang, J.: Towards optimal bag-of-features for object categorization and semantic video retrieval. In: Proceedings of the ACM International Conference on Image and Video Retrieval, pp. 494–501. ACM, New York (2007)
14. Kiros, R., Zhu, Y., Salakhutdinov, R.R., Zemel, R., Urtasun, R., Torralba, A., Fidler, S.: Skip-thought vectors. In: Proceedings of the Advances in Neural Information Processing Systems, pp. 3294–3302. NIPS (2015)
15. Krähenbühl, P., Koltun, V.: Geodesic object proposals. In: Proceedings of the European Conference on Computer Vision, pp. 725–739. Springer, Berlin (2014)

16. Lin, D., Fidler, S., Kong, C., Urtasun, R.: Visual semantic search: retrieving videos via complex textual queries. In: Proceedings of the IEEE Conference on Computer Vision and Pattern Recognition, pp. 2657–2664. IEEE, Piscataway (2014)
17. Lin, T., Zhao, X., Shou, Z.: Single shot temporal action detection. In: Proceedings of the ACM International Conference on Multimedia, pp. 988–996. ACM, New York (2017)
18. Luo, R., Shakhnarovich, G.: Comprehension-guided referring expressions. In: Proceedings of the IEEE Conference on Computer Vision and Pattern Recognition, pp. 7102–7111. IEEE, Piscataway (2017)
19. Ma, S., Sigal, L., Sclaroff, S.: Learning activity progression in LSTMS for activity detection and early detection. In: Proceedings of the IEEE Conference on Computer Vision and Pattern Recognition, pp. 1942–1950. IEEE, Piscataway (2016)
20. Mao, J., Huang, J., Toshev, A., Camburu, O., Yuille, A.L., Murphy, K.: Generation and comprehension of unambiguous object descriptions. In: Proceedings of the IEEE Conference on Computer Vision and Pattern Recognition, pp. 11–20. IEEE, Piscataway (2016)
21. Nagaraja, V.K., Morariu, V.I., Davis, L.S.: Modeling context between objects for referring expression understanding. In: Proceedings of the European Conference on Computer Vision, pp. 792–807. Springer, Berlin (2016)
22. Otani, M., Nakashima, Y., Rahtu, E., Heikkila, J., Yokoya, N.: Learning joint representations of videos and sentences with web image search. In: Proceedings of the European Conference on Computer Vision, pp. 651–667. Springer, Berlin (2016)
23. Pennington, J., Socher, R., Manning, C.: Glove: Global vectors for word representation. In: Proceedings of the Conference on Empirical Methods in Natural Language Processing, pp. 1532–1543. ACL (2014)
24. Regneri, M., Rohrbach, M., Wetzel, D., Thater, S., Schiele, B., Pinkal, M.: Grounding action descriptions in videos. Trans. Assoc. Comput. Linguist. **1**, 25–36 (2013)
25. Ronfard, R.: Reading movies: an integrated DVD player for browsing movies and their scripts. In: Proceedings of the ACM International Conference on Multimedia, pp. 740–741. ACM, New York (2004)
26. Shou, Z., Wang, D., Chang, S.F.: Temporal action localization in untrimmed videos via multi-stage CNNs. In: Proceedings of the IEEE Conference on Computer Vision and Pattern Recognition, pp. 1049–1058. IEEE, Piscataway (2016)
27. Simonyan, K., Zisserman, A.: Two-stream convolutional networks for action recognition in videos. In: Proceedings of the Advances in Neural Information Processing Systems, pp. 568–576. NIPS (2014)
28. Simonyan, K., Zisserman, A.: Very deep convolutional networks for large-scale image recognition (2014), pp. 1–14. arXiv:1409.1556
29. Singh, B., Marks, T.K., Jones, M., Tuzel, O., Shao, M.: A multi-stream bi-directional recurrent neural network for fine-grained action detection. In: Proceedings of the IEEE Conference on Computer Vision and Pattern Recognition, pp. 1961–1970. IEEE, Piscataway (2016)
30. Song, Y.C., Naim, I., Al Mamun, A., Kulkarni, K., Singla, P., Luo, J., Gildea, D., Kautz, H.A.: Unsupervised alignment of actions in video with text descriptions. In: International Joint Conference on Artificial Intelligence, pp. 2025–2031. Morgan Kaufmann, Burlington (2016)
31. Sun, C., Shetty, S., Sukthankar, R., Nevatia, R.: Temporal localization of fine-grained actions in videos by domain transfer from web images. In: Proceedings of the ACM International Conference on Multimedia, pp. 371–380. ACM, New York (2015)
32. Tellex, S., Roy, D.: Towards surveillance video search by natural language query. In: Proceedings of the ACM International Conference on Image and Video Retrieval, pp. 1–9. ACM, New York (2009)
33. Torabi, A., Tandon, N., Sigal, L.: Learning language-visual embedding for movie understanding with natural language, pp. 1–13 (2016). arXiv:1609.08124
34. Uijlings, J.R., Van De Sande, K.E., Gevers, T., Smeulders, A.W.: Selective search for object recognition. Int. J. Comput. Vis. **104**(2), 154–171 (2013)

35. Xu, R., Xiong, C., Chen, W., Corso, J.J.: Jointly modeling deep video and compositional text to bridge vision and language in a unified framework. In: Proceedings of the AAAI Conference on Artificial Intelligence, pp. 2346–2352. AAAI (2015)
36. Yan, R., Hauptmann, A.G., Jin, R.: Negative pseudo-relevance feedback in content-based video retrieval. In: Proceedings of the ACM International Conference on Multimedia, pp. 343–346. ACM, New York (2003)
37. Yu, H., Siskind, J.M.: Grounded language learning from video described with sentences. In: Proceedings of the Annual Meeting of the Association for Computational Linguistics, pp. 53–63. ACL (2013)
38. Yu, L., Poirson, P., Yang, S., Berg, A.C., Berg, T.L.: Modeling context in referring expressions. In: Proceedings of the European Conference on Computer Vision, pp. 69–85. Springer, Berlin (2016)
39. Yu, L., Tan, H., Bansal, M., Berg, T.L.: A joint speaker listener-reinforcer model for referring expressions. In: Proceedings of the IEEE Conference on Computer Vision and Pattern Recognition, pp. 7282–7290. IEEE, Piscataway (2017)
40. Yu, L., Lin, Z., Shen, X., Yang, J., Lu, X., Bansal, M., Berg, T.L.: Mattnet: modular attention network for referring expression comprehension, pp. 1–14 (2018). arXiv:1801.08186
41. Zhuang, B., Wu, Q., Shen, C., Reid, I., Hengel, A.v.d.: Parallel attention: a unified framework for visual object discovery through dialogs and queries pp. 1–11 (2017). arXiv:1711.06370
42. Zitnick, C.L., Dollár, P.: Edge boxes: locating object proposals from edges. In: Proceedings of the European Conference on Computer Vision, pp. 391–405. Springer, Berlin (2014)

Chapter 4
Semantic Pruning Video Moment Localization

4.1 Introduction

Untrimmed videos often contain complex scenes and events, where only certain parts of the content are relevant to the given query, while the rest may be irrelevant and undesired by users [1, 17]. As illustrated in Fig. 4.1, consider a video where a bartender invites a customer to taste various drinks. One might be specifically interested in the moment when "the man is tasting the third drink." Thus, localizing specific moments within a long, untrimmed video using natural language queries, i.e., video moment localization, becomes essential [11].

Despite the increasing interest in video moment localization, it remains a highly challenging task due to the following primary reasons:

1. **Effective Semantic Alignment**: Given an untrimmed video, users can pose a variety of queries to locate their desired moments. These queries are typically categorized into two types: summary queries and detailed queries. As shown in Fig. 4.1, there are two example queries (Q1 and Q2) corresponding to moments marked by green and red bounding boxes, respectively. The first query, a summary query, is "two men are tasting different drinks," which aims to localize a moment involving a series of actions and complex interactions. The second query, a detailed query, is "the man is tasting the third drink," designed to locate a specific moment marked by the clue word "third." Clearly, these two types of queries require different visual-semantic alignments. However, existing approaches face challenges in aligning these distinct query types with the video content:

   - Coarse-grained semantic matching (e.g., clip-by-sentence matching) is common in methods like [1, 8], but it fails to capture fine-grained semantic correlations between key frames and temporal words. This results in poor performance for detailed queries like Q2.

M. Liu et al., *Intelligent and Efficient Video Moment Localization*,
https://doi.org/10.1007/978-3-031-87588-5_4

Fig. 4.1 Examples of localizing moments via two queries within an untrimmed video

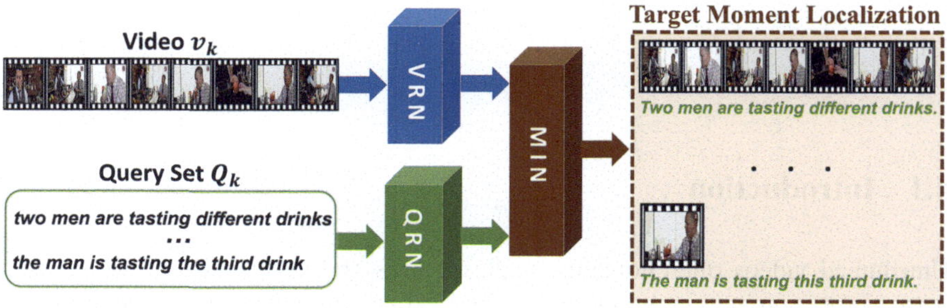

Fig. 4.2 Schematic illustration of our proposed CLEAR model

- Fine-grained semantic interaction (e.g., frame-by-word matching) is adopted by methods like [18, 27, 28], which works well for detailed queries but struggles with summary queries like Q1. These methods tend to overfocus on trivial visual-textual details, diminishing the global understanding required for accurate localization.

2. **Efficient Video Moment Localization**: Localizing the target moments typically requires considering each "video-query" pair and using query-aware attention mechanisms for moment localization. Current methods [15, 17, 18, 27, 28] rely on an iterative "query-for-moment" approach [11], where each query is processed independently, resulting in inefficient localization. Thus, building an efficient video moment localization model remains an open challenge.

To address the aforementioned challenges, we propose a novel end-to-end Coarse-to-fine cross-modaL sEmantic Alignment networRk, named CLEAR, as illustrated in Fig. 4.2. Our approach introduces a dual-path neural network architecture comprising two independent modules: the Video Representation Network (VRN) and the Query Representation Network (QRN). The VRN learns moment representations and models their semantic relevance, while the QRN embeds the queries. We then introduce a Multi-granularity Interaction Network (MIN) for effective cross-modal semantic alignment.

The CLEAR model leverages coarse-grained semantic pruning to filter out irrelevant moments and fine-grained semantic alignment to achieve precise moment localization. This coarse-to-fine approach effectively reduces retrieval overhead and significantly improves localization efficiency. Moreover, we release our code to the public for further research and experimentation.[1]

4.2 Related Work

This work is closely related to the areas of temporal action localization and video moment localization.

Temporal Action Localization Temporal action localization involves predicting the start and end times of specific actions within an untrimmed video. Early works in this domain focused on capturing long-term temporal correlations to enhance fine-grained action detection. Singh et al. [23] proposed a multi-stream Bi-LSTM model that effectively captured these temporal correlations. Similarly, Shou et al. [22] introduced an end-to-end, segment-based 3D Convolutional Neural Network (S-CNN), which extracts spatio-temporal information for improved action localization. However, the sliding window-based proposal generation used in these methods led to increased processing times.

To address this challenge, Gao et al. [9] presented a temporal unit regression network, which accelerats the generation of action proposals and integrats seamlessly with the S-CNN, yielding enhanced performance. Despite these advancements, existing models struggled to fully exploit temporal context information and multi-stream features, leading to limited performance improvements. To overcome these limitations, Chao et al. [4] proposed TAL-Net, a temporal action localization network inspired by Faster R-CNN. TAL-Net aims to better exploit temporal context information and fuse multi-stream features for improved results. Zhao et al. [30] introduced a structural temporal pyramid pooling-based segment network, which outperforms the multi-tower dilated temporal convolution employed in TAL-Net. Although these methods achieve promising performance for simple actions such as "drinking" or "opening a door," they were not designed to handle more complex activities expressed in arbitrary natural language.

Video Moment Localization Video moment localization extends the capabilities of temporal action localization by focusing on identifying specific moments within a video in response to a given query. Gao et al. [8] introduced a temporal coordinate regression network that jointly predicts action proposals and refines temporal boundaries. Concurrently, Hendricks et al. [1] proposed a Moment Context Network (MCN) that integrats both local and global video features for query-guided moment localization. However, early models struggled to effectively capture spatio-temporal interactions across both the visual and textual modalities.

[1] https://github.com/Huyp777/CSUN.

To address these challenges, Liu et al. [17, 18] developed two attention-based networks. The first network focuses on enhancing moment representations by capturing crucial visual context, while the second network extracts relevant keywords from the query. Yuan et al. [27] proposes an Attention-Based Location Regression (ABLR) model that employs a co-attention memory mechanism to capture spatio-temporal interactions between video segments and the query. This model predicts temporal coordinates for the target moment using an attention-based regression network. Xu et al. [26] took a different approach, developing a multi-level model that leveraged video captioning as an auxiliary task to guide temporal boundary prediction.

Later advancements by Lin et al. [15, 28] introduced the Cross-Modal Interaction Network (CMIN), which utilizes graph convolutional networks and multi-head self-attention mechanisms for fine-grained representation learning in each modality. Additionally, a query reconstruction strategy was employed to enhance cross-modal representations. However, CMIN did not fully capitalize on contextual information within the fused features. To address this limitation, Zhang et al. [29] proposed the 2D Temporal Adjacent Network (2D-TAN), which modeled temporal relationships between modalities within a 2D feature map, enabling deeper cross-modal semantic understanding.

Despite the progress made by these approaches, two key limitations persist. First, they fail to incorporate hierarchical similarity evaluation [3] for video-query interactions, limiting their ability to enhance localization accuracy. Second, they lack effective information-pruning strategies to accelerate the localization process. Most of these methods require repeated processing of each query individually to localize multiple moments in a video, leading to inefficient computational performance.

4.3 Methodology

As depicted in Fig. 4.2, our proposed CLEAR model consists of two main components: (1) **Dual-Path Neural Network**: This network includes the Video Representation Network (VRN) and the Query Representation Network (QRN). VRN is designed to generate moment candidates, learn their representations, and model their semantic relevance, while QRN is used to extract representations for the given queries. (2) **Multi-Granularity Interaction Module**: This module is built to achieve cross-modal semantic alignment between the visual and textual modalities. In the following sections, we detail each of these components sequentially.

4.3.1 Problem Formulation

Let $\mathcal{V} = \{v_1, \ldots, v_k, \ldots, v_N\}$ represent a training set of N untrimmed videos, where v_k denotes the k-th video. Each video v_k is annotated with M_k queries, denoted by $Q_k = \{q_{k,1}, \ldots, q_{k,j}, \ldots, q_{k,M_k}\}$, and each query $q_{k,j}$ corresponds to a specific moment with a start time of t_s^j and an end time of t_e^j. Given this training data, our goal is to learn a cross-modal semantic alignment network.

During inference, given a new untrimmed video v_x and its corresponding query set Q_x, we first generate a set of moment candidates, C_x. We then obtain representations for both modalities, denoted as $\widehat{C}_x$ (moments) and $\widehat{Q}_x$ (queries). Subsequently, we use a multi-granularity cross-modal interaction mechanism to effectively localize the target moments.

4.3.2 *Video Representation Network*

To generate the moment candidate set C_k and learn their representations for the given video v_k, we propose a novel video representation network (VRN). This network consists of three main components: enhanced moment representation, position relation determination, and hierarchical semantic modeling.

Enhanced Moment Representing Given an untrimmed video v_k, we first segment it into L video units with a fixed temporal step, and then use a pre-trained 3D convolutional network (C3D) [24] to extract their features, resulting in the feature sequence $\mathbf{U}_k = [u_{k,1}, \ldots, u_{k,l}, \ldots, u_{k,L}]$, where $u_{k,l}$ represents the feature of the l-th video unit. Our goal is to enhance these local features by incorporating contextual information.

To achieve this, we use the Bi-TCN model [2] instead of Bi-LSTM or Bi-GRU [14, 16] to capture long-term semantic dependencies for each unit. As illustrated in Fig. 4.3, each unit feature $u_{k,i}$ is enhanced by integrating contextual information from both directions (pre- and post-context), resulting in a more comprehensive representation. Specifically, inputting $\mathbf{U}_k$ into the Bi-TCN with R layers, the output is formulated as follows:

$$\begin{cases} \widetilde{\mathbf{U}}_k^{(1)} = \theta_1(\mathbf{U}_k, \delta^1, d^1), \\ \quad\vdots \\ \widetilde{\mathbf{U}}_k^{(r)} = \theta_r(\widetilde{\mathbf{U}}_k^{(r-1)}, \delta^r, d^r), \\ \quad\vdots \\ \widetilde{\mathbf{U}}_k^{(R)} = \theta_R(\widetilde{\mathbf{U}}_k^{(R-1)}, \delta^R, d^R), \end{cases} \tag{4.1}$$

where θ_r refers to the 1-D dilated convolution of the r-th layer, as well as d^r and δ^r respectively denote the dilation factor and filter kernel size, respectively.

After obtaining $\widetilde{\mathbf{U}}_k^{(R)}$, we generate moment candidates for localization. As illustrated in Fig. 4.3, we first apply multi-scale temporal pooling [1] to generate representations for all possible moments. For example, the first l unit features are

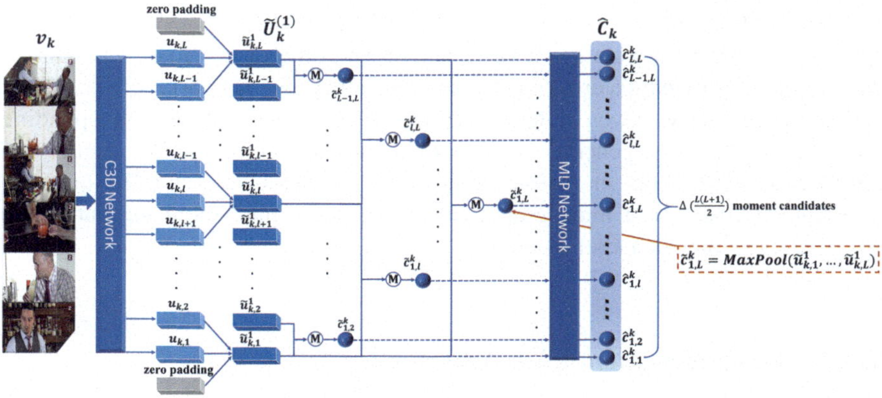

Fig. 4.3 Pipeline of the enhanced moment representation module. The module first uses the C3D model to extract unit features. It then employs a Bi-TCN model to enhance these unit features by capturing pre-context and post-context information, with a kernel size of 3 and appropriate zero padding to maintain consistent output sequence length. Subsequently, a series of multi-scale max pooling operations are applied to generate the basic moment representation set $\widetilde{C}_k$. Finally, an MLP model is used to obtain the enhanced moment representation set $\widehat{C}_k$

max pooled to form the representation for the moment starting at the 1-th unit and ending at the l-th unit, denoted by $\widetilde{c}^k_{1,l} = maxpool(\widetilde{u}_{k,1}, \widetilde{u}_{k,2}, \ldots, \widetilde{u}_{k,l})$. The entire sequence of L units is similarly max-pooled to represent the moment from the first to the L-th unit, denoted by $\widetilde{c}^k_{1,L} = maxpool(\widetilde{u}_{k,1}, \widetilde{u}_{k,2}, \ldots, \widetilde{u}_{k,L})$. In this way, the basic moment representation set C_k, containing representations of $\Delta = \frac{L(1+L)}{2}$ moment candidates, is generated. We then use a multi-layer perception (MLP) network to produce the enhanced moment representation set $\widehat{C}_k \in \mathbb{R}^{\Delta \times d}$ for all moment candidates, where d is the feature dimension of each candidate.

Position Relation Determining To effectively utilize relationships [29] among moment candidates, we project $\widehat{C}_k$ into a temporal-spatial position map, dubbed as $\mathcal{J}_k$. As illustrated in Fig. 4.4, the horizontal and vertical axes of $\mathcal{J}_k$ separately represent the start and duration of each moment. The l-th column of $\mathcal{J}_k$ contains enhanced representations of moment candidates starting at the l-th video unit. Note that, $\mathcal{J}_k \in \mathbb{R}^{L \times L \times d}$ is a lower triangular matrix, where the first two dimensions represent the starting and ending points of each moment. With $\mathcal{J}_k$, the real start and end times of each moment can be easily obtained. Assuming the total duration of video v_k is T_k, the duration of each video unit is $\frac{T_k}{L}$, and the real temporal range of the moment $\widehat{c}^k_{l,l}$ is $\left[(l-1) \times \frac{T_k}{L}, l \times \frac{T_k}{L}\right]$.

Hierarchical Semantic Modeling After obtaining the enhanced moment representations and determining their temporal-spatial relations, we explore the semantic relationship between different moments. In this work, we assume that if two moments have an obvious temporal entailment, they also have an inclusive semantic relationship. For instance, as illustrated in Fig. 4.1, the two moments (highlighted in

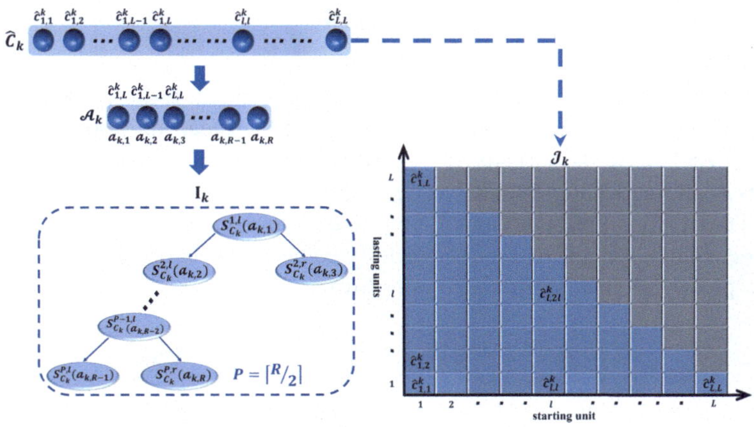

Fig. 4.4 Illustration of hierarchical semantic construction. The process begins by selecting R representative moment candidates from $\widehat{C}_k$ as anchor moments $\mathcal{A}_k$. A video semantic tree $\mathcal{I}_k$ is then built to represent the semantic structure of $\widehat{C}_k$

green and red boxes) have overlapping temporal spans, where the semantic content of the earlier moment ("Two men are tasting different drinks") encompasses the semantic content of the latter one ("The man is tasting this third drink"), as the latter is a specific instance of the former.

To model the hierarchical semantics of $\widehat{C}_k$, we first select R anchor moments $\mathcal{A}_k = \{a_{k,r}\}_{r=1}^{R}$, where each anchor moment $a_{k,r}$ is chosen as follows:

$$a_{k,r} = \begin{cases} \widehat{c}^{k}_{1,L}, & r = 1, \\ \widehat{c}^{k}_{1,L-\frac{r}{2}}, & 1 < r \leq R \,\&\, r \bmod 2 = 0, \\ \widehat{c}^{k}_{L-(\lfloor\frac{r}{2}\rfloor-1),L-(\lfloor\frac{r}{2}\rfloor-1)}, & 1 < r \leq R \,\&\, r \bmod 2 \neq 0. \end{cases} \tag{4.2}$$

Based on $\mathcal{A}_k$, we build a visual semantic tree, dubbed as $\mathcal{I}_k$. This tree divides the overall semantic of v_k into P layers, where $P = \lceil R/2 \rceil$. As shown in Fig. 4.4, the first anchor moment $a_{k,1}$ ($\widehat{c}^{k}_{1,L}$) is treated as the root of $\mathcal{I}_k$, as it covers the semantic of all moment candidates. It is referred to as $\mathcal{S}^{1,l}_{C_k}$. Subsequently, two adjacent anchor moments $a_{k,2}$ ($\widehat{c}^{k}_{1,L-1}$) and $a_{k,3}$ ($\widehat{c}^{k}_{L,L}$) form the left and right semantic nodes of the second layer, denoted as $\mathcal{S}^{2,l}_{C_k}$ and $\mathcal{S}^{2,r}_{C_k}$, respectively. The semantic node $\mathcal{S}^{2,l}_{C_k}$ covers the semantic of moments $\widehat{c}^{k}_{i,j}$ $(1 \leq i \leq j \leq L-1)$. Analogously, $a_{k,R-1}$ and $a_{k,R}$ serve as semantic nodes of the P-th layer. For each pair of adjacent semantic layers, $\mathcal{S}^{p,l}_{C_k}$ and $\mathcal{S}^{p,r}_{C_k}$ are subsets of $\mathcal{S}^{p-1,l}_{C_k}$, i.e., $\mathcal{S}^{p,l}_{C_k} \subset \mathcal{S}^{p-1,l}_{C_k}$ and $\mathcal{S}^{p,r}_{C_k} \subset \mathcal{S}^{p-1,l}_{C_k}$ $(1 \leq p \leq P)$.

4.3.3 Query Representation Network

To learn query representations, we use the pre-trained Bidirectional Encoder Representations from Transformers (BERT) [7, 19] instead of word embedding-based Bi-LSTM or its variants [5, 6]. BERT provides corresponding semantic features to all queries, represented as $\widetilde{Q}_k = \{\widetilde{q}_{k,1}, \ldots, \widetilde{q}_{k,j}, \ldots, \widetilde{q}_{k,M_k}\}$, where $\widetilde{q}_{k,j}$ denotes the feature representation of the j-th query related to video v_k.

Subsequently, we utilize a one-layer fully connected network to obtain the final query representation $\widehat{Q}_k$, formulated as follows:

$$\widehat{Q}_k = \sigma_q(\mathbf{W}_q\widetilde{Q}_k + \mathbf{b}_q) = \{\widehat{q}_1^k, \ldots, \widehat{q}_j^k, \ldots, \widehat{q}_{M_k}^k\}, \tag{4.3}$$

where σ_q represents the ReLU activation function [25], and $\mathbf{W}_q$ and $\mathbf{b}_q$ are the weight matrix and bias vector,respectively. $\widehat{q}_j^k$ refers to the final representation of the j-th query related to video v_k.

4.3.4 Multi-Granularity Interaction Network

After obtaining representations $\widehat{C}_k$ and $\widehat{Q}_k$, we develop the MIN to explore cross-modal semantic correlation.

Coarse-Grained Semantic Alignment We project anchor moments $\mathcal{A}_k \subset \widehat{C}_k$ and the corresponding query representations $\widehat{Q}_k$ into a shared isomorphic representation space for semantic similarity metric learning. As shown in Fig. 4.5a, we utilize Intersection over Union (IoU) [10] to calculate the cross-modal semantic similarity of each "*moment-query*" pair. Specifically, the IoU between a moment and its corresponding query is computed as follows:

$$IoU_{\widehat{q}_{k,j}}^{a_{k,p}} = \frac{\min\left(t_e^j, t_e^p\right) - \max\left(t_s^j, t_s^p\right)}{\max\left(t_e^j, t_e^p\right) - \min\left(t_s^j, t_s^p\right)}, \tag{4.4}$$

where t_s^p and t_e^p denote the start and end points of the p-th anchor moment $a_{k,p}$, and t_s^j and t_e^j represent the start and end points of the target moment depicted by the j-th query $\widehat{q}_{k,j}$.

Based on the IoU scores of the "*moment-query*" pairs, we construct a similarity matrix $\mathcal{M}_k^*$ to preserve their intrinsic semantic similarities. We then propose a loss function Γ_1 for semantic similarity metric learning:

$$\Gamma_1 = \sum_k \left(\|\mathcal{A}_k^T\widehat{Q}_k - d\mathcal{M}_k^*\|_F^2 \right), \tag{4.5}$$

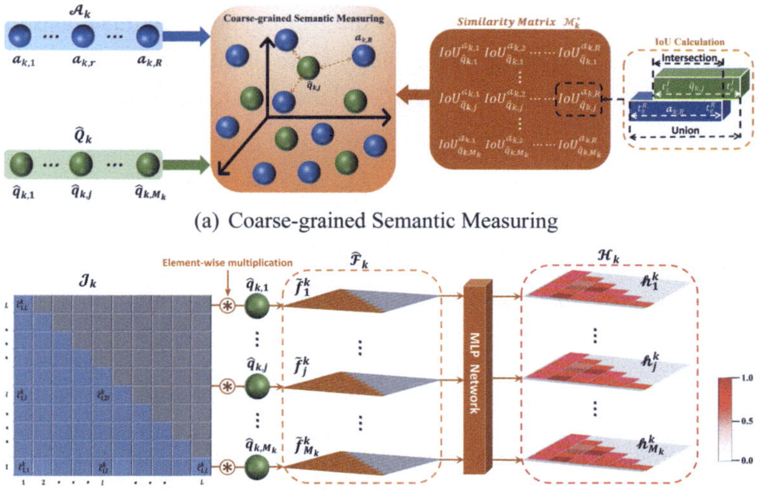

(a) Coarse-grained Semantic Measuring

(b) Fine-grained Semantic Fusing

Fig. 4.5 Illustration of the Multi-granularity Interaction Network (MIN). (**a**) Coarse-grained semantic measurement between anchor moments $\mathcal{A}_k$ and query representations $\widehat{Q}_k$ for semantic similarity metric learning. (**b**) Fine-grained semantic fusion between $\mathcal{J}_k$ and $\widehat{Q}_k$ to create the cross-modal representation $\widehat{\mathcal{F}}_k$ and compute the confidence map $\mathcal{H}_k$

where $\|\cdot\|_F$ denotes the Frobenius norm, d represents the feature dimension of multi-modal features (i.e., anchor moment and query), and $\mathcal{M}_k^*$ is the cross-modal similarity matrix.

Fine-Grained Semantic Alignment We adopt element-wise multiplication $\circledast$ to fuse the multi-modal features. As shown in Fig. 4.5b, the moment based temporal-spatial position map $\mathcal{J}_k$ is successively fused with each $\widehat{q}_j^k \in \widehat{Q}_k$ to form the corresponding cross-modal representation map $\widehat{\mathcal{F}}_k = \{\widetilde{f}_j^k\}_{j=1}^{M_k}$. Specifically, for each $\widetilde{f}_j^k \in \widehat{\mathcal{F}}_k$, it is calculated as follows:

$$\widetilde{f}_j^k = \left[\widehat{q}_j^k \circledast \widehat{c}_{m,n}^k\right]_{1\leq m\leq n}^{L,L}, \tag{4.6}$$

where $\circledast$ denotes the element-wise multiplication.

Using $\widehat{\mathcal{F}}_k$, the confidence map set ($\mathcal{H}_k$) can be obtained by an MLP network. In Fig. 4.5b, each $\hbar_j^k \in \mathcal{H}_k$ indicates the matching scores between the j-th query $\widehat{q}_j^k$ and Δ moment candidates in $\mathcal{J}_k$. Specifically, the darker the color, the greater the semantic similarity. To generate confidence scores more accurately, we use the corresponding IoU values as supervision information and propose a binary cross entropy loss function for semantic fusing learning between $\mathcal{J}_k$ and $\widehat{Q}_k$:

$$\Phi_k = \sum_{j=1}^{M_k} \frac{1}{\Delta} \sum_{i=1}^{\Delta} \left[o_i^j \log p_i^j + \left(1 - o_i^j\right) \log \left(1 - p_i^j\right)\right], \tag{4.7}$$

where o_i^j denotes the IoU value between the i-th moment candidate and the j-th query, and p_i^j is the corresponding similarity score of them in the confidence map $\hbar_j^k$. The overall loss function for N videos and their related queries is formulated as follows:

$$\Gamma_2 = \sum_{k=1}^{N} \Phi_k. \tag{4.8}$$

The final objective function of our CLEAR model is a combination of the two loss functions (Γ_1 and Γ_2):

$$\Psi = \Gamma_1 + \alpha \Gamma_2, \tag{4.9}$$

where α is a non-negative trade-off parameter. The multi-granularity (coarse and fine-grained) semantic alignment is thereby integrated into a unified cross-modal learning framework.

4.3.5 *Inference*

Upon completion of the model training, the proposed CLEAR model leverages the learned cross-modal semantic representations for effective moment localization.

Given an untrimmed video v_x and its corresponding query set Q_x, the first step involves encoding the video and queries using the VRN and QRN, respectively. Next, the MIN is employed to compute the semantic similarity between the query representations $\widehat{Q}_x$ and the semantic tree of the video, denoted as $\mathcal{I}_x$, while simultaneously generating the confidence score map $\mathcal{H}_x$. The target moments are then localized by aligning the semantic information in a coarse-to-fine manner with respect to the given queries. A detailed description of this process is provided in Algorithm 1.

For instance, consider the query $\widehat{q}_2^x$ from Q_x, as depicted in Fig. 4.6. Upon evaluating the semantic similarity between $\widehat{q}_2^x$ (highlighted in yellow) and the three anchor moments $a_{x,1}$, $a_{x,2}$, and $a_{x,3}$ (highlighted in blue), the following observations are made: (1) In compared to $a_{x,1}$, $\widehat{q}_2^x$ exhibits a higher semantic similarity with both $a_{x,2}$ and $a_{x,3}$. Consequently, the search is directed towards identifying a more suitable semantic node from the second layer of the video semantic tree $\widehat{\mathcal{I}}_x$. (2) Among the anchor moments in the second layer, $\widehat{q}_2^x$ is most semantically aligned with $a_{x,2}$. Thus, the refined semantic node $\mathcal{S}_{C_x}^{2,l}$ from $\widehat{\mathcal{I}}_x$ is selected, as indicated in line 13 of Algorithm 1. Based on the guidance of this selected semantic node $\mathcal{S}_{C_x}^{2,l}$, the target moment is localized within the yellow-highlighted region of the confidence map $\hbar_2^x$.

Algorithm 1: Coarse-to-fine moment localization

Input: $\widehat{Q}_x$, $\mathcal{I}_x$ and $\mathcal{H}_x$.

Output: Target moments corresponding to Q_x.

Initialize: Set intra-layer similarity threshold μ, ρ ($\mu < \rho$); Set P to the total layer of $\mathcal{I}_x$;

for *each* $\widehat{q}_j^x$ *in* $\widehat{Q}_x$ **do**

  i = 1; cur = $\mathcal{S}_{C_x}^{1,l}$; //Set the current semantic node to the root of $\mathcal{I}_x$;

  while $i < P$ **do**

    x = ED($\widehat{q}_j^x$, cur); // Calculate the Euclidean distance between query $\widehat{q}_j^x$ and current semantic node;

    y = ED($\widehat{q}_j^x$, cur.ln); // Calculate the Euclidean distance between query $\widehat{q}_j^x$ and the left semantic node of cur;

    z = ED($\widehat{q}_j^x$, cur.rn);

    if $Max(y, z) < x$ **then**

      $\eta_j = \frac{y-z}{y}$; // Judge the intra-layer similarity;

      if $\eta_j > \rho$ **then**

        cur = $\mathcal{S}_{C_x}^{i+1,r}$ and break;

      else if $\eta_j \leq \mu$ **then**

        cur = $\mathcal{S}_{C_x}^{i+1,l}$;

        i = i+1;

      else

        cur = $\mathcal{S}_{C_x}^{i,l} - \mathcal{S}_{C_x}^{i+1,l} - \mathcal{S}_{C_x}^{i+1,r}$ and break;

    else

      cur = $\mathcal{S}_{C_x}^{i,l}$ and break;

  Obtain the matching scores from $\hbar_j^x$ related to the refined semantic node: *cur*;

  Output the target moment of $\widehat{q}_j^x$ by ranking matching scores;

Similarly, the target moments for other queries can be localized effectively by concentrating on the relatively smaller areas highlighted within their respective confidence maps. This coarse-to-fine strategy facilitates both efficient and precise localization of the relevant video moments.

4.4 Experiments

We first evaluated the effectiveness of our proposed CLEAR model on two temporal moment localization datasets: ActivityNet Captions and TACoS. We then investigated the contribution of each component of CLEAR to localization performance through ablation studies. Finally, we conducted an efficiency analysis and qualitative comparison to provide further insights into the strengths of our model.

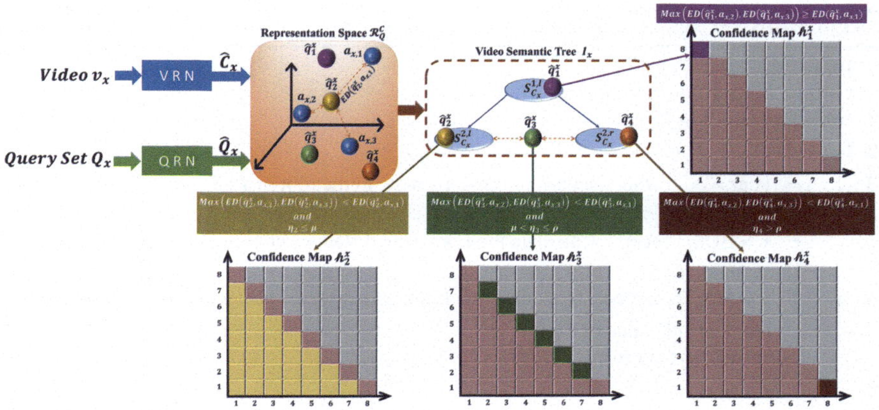

Fig. 4.6 A simple illustration of our inference process. VRN first divides the given v_x into 8 equal video units, and then obtains the enhanced moment representation set $\widehat{C}_x$ and the two-layer semantic tree $\mathcal{I}_x$. Meanwhile, QRN encodes queries into $\widehat{Q}_x$. Thereafter, the cross-modal semantic similarity and corresponding confidence score maps $\mathcal{H}_x$ can be obtained by MIN. According to Algorithm 1, we can determine the appropriate range in $\mathcal{I}_x$ and obtain the refined $\hbar_j^x$ for moment localization

4.4.1 *Experimental Settings*

4.4.1.1 Datasets

We evaluated our proposed CLEAR model on two benchmark datasets: ActivityNet Captions and TACoS.

ActivityNet Captions [20] It comprises 20,000 untrimmed videos paired with over 70,000 natural language queries and temporal annotations, covering a wide range of human activities such as eating, drinking, and socializing. The average video and moment durations are 117.74 seconds and 37.14 seconds, respectively, with the average query length being 13 words.

TACoS [8] It is derived from the MPII Cooking Composite Activities dataset [21], containing 127 untrimmed cooking videos and 18,000 natural language queries. The average video duration is 296.21 seconds, with target moments averaging 6.10 seconds. While ActivityNet Captions provides a diverse range of video content across various domains, TACoS focuses specifically on cooking activities. The latter presents additional challenges for moment localization, as each video typically involves a higher number of target moments that need to be localized.

4.4.1.2 Evaluation Metrics

To comprehensively evaluate the performance of both our model and the baseline methods, we utilized the "R@n, IoU=m" metric proposed by Hu et al. [10]. This metric, denoted as $R(n, m)$, evaluates the top-n retrieved moments, where $n \in \{1, 5\}$ represents the number of retrieved moments, and $m \in \{0.3, 0.5, 0.7\}$ corresponds to the Intersection over Union (IoU) threshold. In addition to this, we measured the total localization time ($\mathcal{R}_T$) and average localization time ($\mathcal{R}_A$) to assess the efficiency of the localization process.

4.4.1.3 Implementation Details

For each video in the datasets, we partitioned the video into units, each consisting of 16 consecutive frames, with an overlap of 8 frames between adjacent units. Features for these video units were extracted using a pre-trained C3D model [24]. Specifically, 500-dimensional features were used for the ActivityNet Captions dataset, while 4096-dimensional features were employed for the TACoS dataset. All parameters of the CLEAR model were initialized randomly, and we utilized the Adam optimizer [12, 13] to minimize the multi-task loss function.

In our experiments, we configured the semantic tree to consist of two layers (i.e., $P = 2$) for the ActivityNet Captions dataset and four layers (i.e., $P = 4$) for the TACoS dataset. The hyperparameter α was set to 1 across all experiments. During training, the batch size and learning rate were set to 32 and 0.0005, respectively, for ActivityNet Captions, and 16 and 0.001, respectively, for TACoS. The maximum number of epochs was set to 200 empirically, and an early stopping criterion was employed to ensure model convergence.

All experiments were conducted on a workstation running Ubuntu 16.04.6 LTS, equipped with an Intel Xeon E7 CPU, 1 TB of memory, and four NVIDIA RTX 2080Ti GPUs.

4.4.2 Performance Comaprison

To demonstrate the effectiveness of our proposed CLEAR model, we compared it with seven state-of-the-art baselines:[2] MCN [1], ACRN [17], CTRL [8], ABLR [27], CMIN [15], QSPN [26], and 2D-TAN [29].

---

[2] ABLR directly regresses the temporal coordinates based on maximum co-attention weights/features, and therefore only generates R@1 results. Moreover, we adopted the latest extended version of CMIN [15]. In addition, there are two variants of the 2D-TAN model: 2D-TAN Pool and 2D-TAN Conv. To ensure a fair comparison, we used the average $R(n, m)$ of these variants as the overall performance of 2D-TAN.

Table 4.1 Localization accuracy comparison between our proposed model and several state-of-the-art baselines on ActivityNet Captions dataset (p-value*: p-value over $R(1, 0.5)$)

Method	R@1 IoU=0.7	R@1 IoU=0.5	R@1 IoU=0.3	R@5 IoU=0.7	R@5 IoU=0.5	R@5 IoU=0.3	p-value*
MCN	6.43%	21.36%	39.35%	29.70%	53.23%	68.12%	1.28E-32
CTRL	10.34%	29.01%	47.43%	37.54%	59.17%	75.32%	1.90E-29
ACRN	11.25%	31.67%	49.70%	38.57%	60.34%	76.50%	5.58E-28
QSPN	13.43%	33.26%	52.13%	40.78%	62.39%	77.72%	5.85E-27
ABLR	15.71%	36.79%	55.67%	–	–	–	4.14E-24
CMIN	24.48%	44.62%	**64.41%**	52.96%	69.66%	82.39%	2.40E-05
2D-TAN	26.96%	44.28%	59.10%	62.11%	76.89%	85.59%	3.77E-06
CLEAR	**27.05%**	**44.73%**	59.76%	**62.13%**	**77.06%**	**85.63%**	–

The best experimental results are highlighted in bold

Table 4.2 Localization accuracy comparison between our proposed model and several state-of-the-art baselines on TACoS dataset (p-value*: p-value over $R(1, 0.5)$)

Method	R@1 IoU=0.7	R@1 IoU=0.5	R@1 IoU=0.3	R@5 IoU=0.7	R@5 IoU=0.5	R@5 IoU=0.3	p-value*
MCN	1.00%	1.25%	1.64%	1.01%	1.25%	2.03%	7.12E-29
CTRL	6.98%	13.08%	18.17%	15.60%	26.13%	35.97%	6.78E-25
ACRN	8.28%	14.11%	20.06%	15.43%	26.30%	38.11%	1.94E-24
QSPN	7.56%	15.23%	20.15%	14.35%	25.30%	36.72%	6.55E-24
ABLR	6.13%	9.30%	18.90%	–	–	–	2.19E-26
CMIN	10.21%	19.57%	27.33%	18.16%	28.53%	43.35%	1.79E-21
2D-TAN	12.47%	25.23%	36.26%	25.27%	44.64%	57.40%	1.72E-16
CLEAR	**14.34%**	**25.61%**	**40.18%**	**26.05%**	**50.16%**	**63.21%**	–

The best experimental results are highlighted in bold

Tables 4.1 and 4.2 show the overall localization accuracy results for CLEAR and the baselines on the ActivityNet Captions and TACoS datasets, respectively. Based on these results, we observed the following:

- Both MCN and CTRL demonstrate lower performance compared to other models, as they incorporate contextual information either by roughly fusing the entire video features or by extending moment boundaries within a limited scale. Specifically, MCN tends to fuse excessive contextual information, potentially introducing noise and degrading the discriminative power of context representation. Meanwhile, CTRL only considers limited pre- and post-moment extensions, failing to capture long-term dependencies effectively.
- ACRN, QSPN, and ABLR outperform MCN and CTRL, reflecting the benefit of attention mechanisms in highlighting crucial modality information, thereby enhancing feature representation and improving localization performance.
- CMIN and 2D-TAN achieve relatively higher localization accuracy compared to other baselines. This highlights the importance of multi-stage cross-modal interaction and long-range semantic dependency modeling in enhancing localization accuracy.

- CLEAR delivers the highest localization accuracy, outperforming all state-of-the-art baselines, particularly on the TACoS dataset. While TACoS presents greater challenges for moment localization, CLEAR achieves the best performance, validating the effectiveness of our coarse-to-fine semantic alignment approach.

Furthermore, we conducted significance tests between CLEAR and each of the baselines for the $R(1, 0.5)$ metric, using results from 20 independent rounds. The p-values for all comparisons were found to be less than 0.05, indicating that the superiority of our CLEAR model over the baselines is statistically significant.

4.4.3 Ablation Study

We conducted ablation studies on our proposed model by removing specific components to generate the following ablation models:

- **CLEAR-V**: Removed the Bi-TCN from the VRN, directly using feature pooling for video encoding.
- **CLEAR-C**: Eliminated the fine-grained component from MIN, relying solely on coarse-grained semantic alignment, i.e., semantic similarity evaluation in the shared isomorphic representation space for moment localization.
- **CLEAR-F**: Removed the coarse-grained component, utilizing only fine-grained semantic alignment to sequentially retrieve the highest similarity score in the entire confidence map for moment localization.

We evaluated these ablation models on the ActivityNet Captions and TACoS datasets. The results are presented in Fig. 4.7. Based on these results, we make the following observations:

- CLEAR-C exhibits the worst localization accuracy across both datasets, particularly on the more challenging TACoS dataset. This suggests that relying solely on coarse-grained semantic metrics is insufficient for capturing complex semantic similarities, leading to ineffective moment localization. However, since CLEAR-C omits the time-consuming fine-grained semantic matching, it is more efficient compared to the other variants.
- CLEAR-V shows relatively low accuracy but higher efficiency. On the one hand, removing Bi-TCN improves processing efficiency. On the other hand, omitting the long-term dependency modeling for video encoding adversely impacts accuracy, particularly in more challenging scenarios. For instance, as seen in Fig. 4.7b, when IoU = 0.7 on TACoS, the accuracy degrades significantly.
- CLEAR-F achieves relatively high accuracy but has the lowest efficiency, indicating that the absence of coarse-grained semantic pruning hinders overall localization efficiency improvement.
- CLEAR achieves a good balance between localization accuracy and efficiency on both datasets. The results demonstrate that CLEAR attains the best performance trade-off, verifying that the Bi-TCN, coarse-grained, and fine-grained semantic alignment components all contribute positively to moment localization.

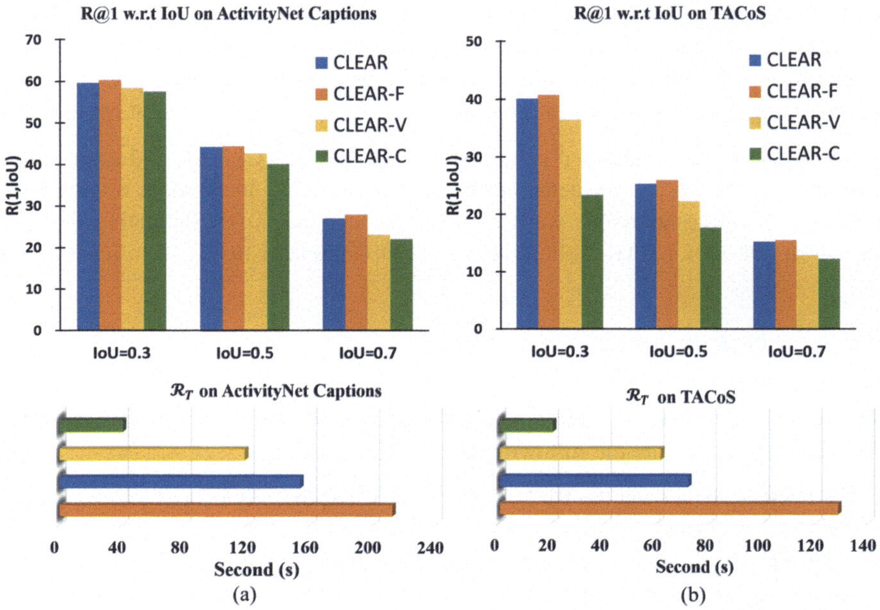

Fig. 4.7 Comparison of localization accuracy and efficiency between our CLEAR model and its variants on the ActivityNet Captions and TACoS datasets. (**a**) Results based on R@1 for IoU values of {0.3, 0.5, 0.7}, along with $\mathcal{R}_T$, on the ActivityNet Captions dataset. (**b**) Results based on R@1 for IoU values of {0.3, 0.5, 0.7}, along with $\mathcal{R}_T$, on the TACoS dataset

Table 4.3 Running time comparison between our proposed CLEAR and the state-of-the-art baselines on ActivityNet Captions and TACoS datasets

Method	ActivityNet captions		TACoS	
	$\mathcal{R}_A$ (s)	$\mathcal{R}_T$ (s)	$\mathcal{R}_A$ (s)	$\mathcal{R}_T$ (s)
MCN	5.23	89,021.56	11.53	47,095.03
CTRL	2.57	43,715.49	4.90	19,989.97
ACRN	3.31	56,425.11	6.14	25,087.73
ABLR	0.04	765.98	0.18	717.68
CMIN	0.67E-2	114.51	0.96E-2	39.02
2D-TAN	1.45E-2	247.43	0.53E-1	217.57
CLEAR	0.90E-2	153.56	1.75E-2	71.59

4.4.4 Efficiency Analysis

Table 4.3 presents a comparison of localization efficiency between our CLEAR model and the baselines on the ActivityNet Captions and TACoS datasets. Analyzing the results in Table 4.3, we observe the following:

- MCN, CTRL, and ACRN demonstrate relatively low efficiency. This is mainly due to their reliance on densely sliding window-based moment candidate generation and iterative retrieval for locating the target moments, which makes the localization process time-consuming.

- ABLR and 2D-TAN achieve better efficiency ($\mathcal{R}_A$ and $\mathcal{R}_T$) compared to the above models. This can be attributed to: (1) ABLR divides the video evenly into moment candidates and directly performs temporal coordinate regression, reducing computational cost. (2) 2D-TAN uses a sparse sampling strategy to efficiently model cross-domain feature correlations.
- CMIN has relatively higher efficiency than other baselines, mainly due to its use of PCA and downsampling to reduce the feature dimensions and length, which accelerates cross-modal interactions and overall efficiency.
- CLEAR demonstrates better efficiency than all baselines except CMIN. This may be because CLEAR does not apply feature dimensionality reduction during representation encoding, and its multi-granularity semantic alignment introduces additional computational overhead. However, when considering the ablation model CLEAR-C, which achieves higher efficiency than CMIN (as shown in Fig. 4.7 and Table 4.3), we can conclude that CLEAR strikes a strong balance between high accuracy and efficiency.

4.4.5 Qualitative Results

To qualitatively validate the effectiveness of our CLEAR model, we compared it with the two best-performing baselines (CMIN and 2D-TAN) using examples from the ActivityNet Captions and TACoS datasets. As shown in Fig. 4.8, CLEAR

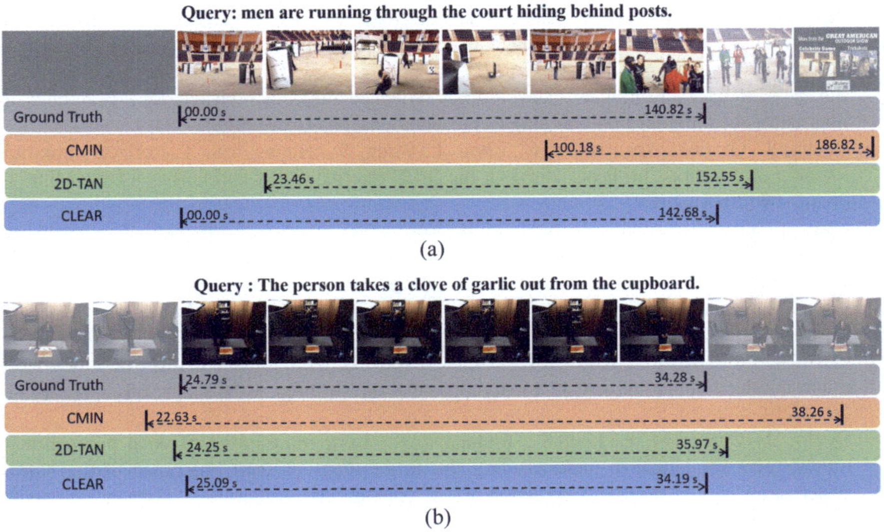

Fig. 4.8 Moment localization results on ActivityNet Captions and TACoS datasets. All the above figures are the R@1 results. (**a**) Moment localization results on ActivityNet captions. (**b**) Moment localization results on TACoS

achieves more accurate localization results, with the target moments returned by our model showing a larger IoU with the ground truth compared to the baselines. The improved results can be attributed to two main factors: (1) CLEAR captures hierarchical semantic dependencies and temporal-spatial relationships more effectively than CMIN and 2D-TAN. And (2) the use of multi-granularity semantic alignment for cross-modal correlation modeling significantly enhances the localization performance.

4.5 Summary

In this chapter, we present CLEAR, a deep cross-modal semantic alignment network for video moment localization. Specifically, we designed a dual-path neural network to model the semantic information of video moments and to understand query intentions. We then developed a multi-granularity interaction module to achieve effective cross-modal semantic alignment, enabling accurate localization of target moments. To evaluate the performance of CLEAR, we conducted extensive experiments on two public benchmark datasets, demonstrating the superiority of our model compared to several state-of-the-art approaches.

References

1. Anne Hendricks, L., Wang, O., Shechtman, E., Sivic, J., Darrell, T., Russell, B.: Localizing moments in video with natural language. In: Proceedings of the IEEE International Conference on Computer Vision, pp. 5803–5812 (2017)
2. Bai, S., Kolter, J.Z., Koltun, V.: Convolutional sequence modeling revisited. In: Proceedings of International Conference on Learning Representations, pp. 1–14 (2018)
3. Chang, S., Qi, G.J., Aggarwal, C.C., Zhou, J., Wang, M., Huang, T.S.: Factorized similarity learning in networks. In: Proceedings of the IEEE International Conference on Data Mining, pp. 60–69 (2014)
4. Chao, Y., Vijayanarasimhan, S., Seybold, B., Ross, D.A., Deng, J., Sukthankar, R.: Rethinking the faster R-CNN architecture for temporal action localization. In: Proceedings of the IEEE Conference on Computer Vision and Pattern Recognition, pp. 1130–1139 (2018)
5. Chen, X., Song, X., Cui, S., Gan, T., Cheng, Z., Nie, L.: User identity linkage across social media via attentive time-aware user modeling. IEEE Trans. Multimedia **23**, 3957–3967 (2020)
6. Chen, X., Song, X., Ren, R., Zhu, L., Cheng, Z., Nie, L.: Fine-grained privacy detection with graph-regularized hierarchical attentive representation learning. ACM Trans. Inf. Syst. **38**, 1–26 (2020)
7. Devlin, J., Chang, M.W., Lee, K., Toutanova, K.: Bert: pre-training of deep bidirectional transformers for language understanding. In: Proceedings of the Conference of the North American Chapter of the Association for Computational Linguistics: Human Language Technologies, pp. 4171–4186 (2019)
8. Gao, J., Sun, C., Yang, Z., Nevatia, R.: Tall: temporal activity localization via language query. In: Proceedings of the IEEE International Conference on Computer Vision, pp. 5267–5275 (2017)

9. Gao, J., Yang, Z., Sun, C., Chen, K., Nevatia, R.: Turn tap: temporal unit regression network for temporal action proposals. In: Proceedings of the IEEE International Conference on Computer Vision, pp. 3628–3636 (2017)
10. Hu, R., Xu, H., Rohrbach, M., Feng, J., Saenko, K., Darrell, T.: Natural language object retrieval. In: Proceedings of the IEEE Conference on Computer Vision and Pattern Recognition, pp. 4555–4564 (2016)
11. Hu, Y., Liu, M., Su, X., Gao, Z., Nie, L.: Video moment localization via deep cross-modal hashing. IEEE Trans. Image Process. **30**, 4667–4677 (2021)
12. Hu, Y., Zhan, P., Xu, Y., Zhao, J., Li, Y., Li, X.: Temporal representation learning for time series classification. Neural Comput. Appl. **33**, 3169–3182 (2021)
13. John, D., Elad, H., Yoram, S.: Adaptive subgradient methods for online learning and stochastic optimization. J. Mach. Learn. Res. **12**, 2121–2159 (2011)
14. Li, Y., Liu, M., Yin, J., Cui, C., Xu, X.S., Nie, L.: Routing micro-videos via a temporal graph-guided recommendation system. In: Proceedings of the ACM International Conference on Multimedia, pp. 1464–1472 (2019)
15. Lin, Z., Zhao, Z., Zhang, Z., Zhang, Z., Cai, D.: Moment retrieval via cross-modal interaction networks with query reconstruction. IEEE Trans. Image Process. **29**, 3750–3762 (2020)
16. Liu, M., Nie, L., Wang, M., Chen, B.: Towards micro-video understanding by joint sequential-sparse modeling. In: Proceedings of the ACM International Conference on Multimedia, pp. 970–978 (2017)
17. Liu, M., Wang, X., Nie, L., He, X., Chen, B., Chua, T.S.: Attentive moment retrieval in videos. In: Proceedings of the ACM SIGIR Conference on Research and Development in Information Retrieval, pp. 15–24 (2018)
18. Liu, M., Wang, X., Nie, L., Tian, Q., Chen, B., Chua, T.: Cross-modal moment localization in videos. In: Proceedings of the ACM International Conference on Multimedia, pp. 843–851 (2018)
19. Qu, L., Liu, M., Cao, D., Nie, L., Tian, Q.: Context-aware multi-view summarization network for image-text matching. In: Proceedings of the ACM International Conference on Multimedia, pp. 1047–1055 (2020)
20. Ranjay, K., Kenji, H., Frederic, R., Li, F.F., Juan, C.N.: Dense-captioning events in videos. In: Proceedings of the IEEE International Conference on Computer Vision, pp. 706–715 (2017)
21. Rohrbach, M., Regneri, M., Andriluka, M., Amin, S., Pinkal, M., Schiele, B.: Script data for attribute-based recognition of composite activities. In: Proceedings of the European Conference on Computer Vision, pp. 144–157 (2012)
22. Shou, Z., Wang, D., Chang, S.F.: Temporal action localization in untrimmed videos via multi-stage CNNs. In: Proceedings of the IEEE Conference on Computer Vision and Pattern Recognition, pp. 1049–1058 (2016)
23. Singh, B., Marks, T.K., Jones, M., Tuzel, O., Shao, M.: A multi-stream bi-directional recurrent neural network for fine-grained action detection. In: Proceedings of the IEEE Conference on Computer Vision and Pattern Recognition, pp. 1961–1970 (2016)
24. Tran, D., Bourdev, L., Fergus, R., Torresani, L., Paluri, M.: Learning spatiotemporal features with 3d convolutional networks. In: Proceedings of the IEEE International Conference on Computer Vision, pp. 4489–4497 (2015)
25. Xu, B., Wang, N., Chen, T., Li, M.: Empirical evaluation of rectified activations in convolutional network. In: Proceedings of the International Conference on Machine Learning, pp. 1–5 (2015)
26. Xu, H., He, K., Plummer, B.A., Sigal, L., Sclaroff, S., Saenko, K.: Multilevel language and vision integration for text-to-clip retrieval. In: Proceedings of the AAAI Conference on Artificial Intelligence, pp. 9062–9069 (2019)
27. Yuan, Y., Mei, T., Zhu, W.: To find where you talk: temporal sentence localization in video with attention based location regression. In: Proceedings of the AAAI Conference on Artificial Intelligence, pp. 9159–9166 (2019)

28. Zhang, Z., Lin, Z., Zhao, Z., Xiao, Z.: Cross-modal interaction networks for query-based moment retrieval in videos. In: Proceedings of the ACM SIGIR Conference on Research and Development in Information Retrieval, pp. 655–664 (2019)
29. Zhang, S., Peng, H., Fu, J., Luo, J.: Learning 2d temporal adjacent networks for moment localization with natural language. In: Proceedings of the AAAI Conference on Artificial Intelligence, pp. 1–9 (2020)
30. Zhao, Y., Xiong, Y., Wang, L., Wu, Z., Tang, X., Lin, D.: Temporal action detection with structured segment networks. Int. J. Comput. Vis. **128**, 74–95 (2020)

Chapter 5
Semantic Collaborative Video Moment Localization

5.1 Introduction

As illustrated in Fig. 5.1, surveillance video captures the ongoing activities in a parking lot, but a police officer may only be interested in the specific moment, "The man raised a gun and then hijacked the cleaner into the 7–11 store", which is pivotal for criminal evidence verification. Evidently, compared to manually searching for the target moment, automatically localizing the desired video segment through a language-based query is not only efficient but also essential for practical use.

Despite its promising potential, video moment localization is inherently challenging due to the following two aspects:

- **Intra-modal Semantic Modeling**: As depicted in Fig. 5.1, temporal and spatial semantic cues, such as "raised a gun", are often present within the textual query, and complex events and scenes are usually captured in untrimmed videos. Therefore, an efficient semantic understanding of both textual and visual modalities is indispensable. To address this need, substantial research has focused on improving intra-modal semantic modeling. Specifically, most existing approaches [17, 34, 40] employ bidirectional long short-term memory networks (Bi-LSTM) or their variants for semantic representation of each modality. However, these methods often struggle to capture long-term dependencies due to limited memory capabilities, particularly in long and complex video and query sequences, leading to suboptimal accuracy. Consequently, robust intra-modal semantic modeling remains crucial for effective video moment localization.
- **Inter-modal Semantic Collaboration**: Video moment localization requires a comprehensive understanding of the correlations between textual and visual modalities. However, previous studies have primarily concentrated on modeling semantic consistency between the two modalities [1, 5, 34], largely neglecting the significance of semantic complementarity—the exclusive information presented

M. Liu et al., *Intelligent and Efficient Video Moment Localization*,
https://doi.org/10.1007/978-3-031-87588-5_5

Fig. 5.1 Video moment localization on a real-world surveillance video. A proper understanding of semantic consistency and complementary information (highlighted in blue and orange, respectively) is critical for enhancing cross-modal semantic reasoning capabilities. Specifically, semantically consistent cues, such as "raised a gun" presented in both modalities, must be aligned. Complementary information, such as "7–11 store" which is clear in the query but absent in the video, should be supplemented to aid in understanding unclear visual content, thereby improving localization accuracy

in one modality but absent in the other [31]. As illustrated in Fig. 5.1, textual information such as "into the 7–11 store" is crucial for identifying the end timestamp of the target moment, particularly when the visual representation is unclear. Therefore, supplementing the semantic information from one modality to enhance the collaborative understanding of both modalities is essential. The effective realization of semantic collaboration (consistency and complementarity) between modalities remains a largely unresolved issue in video moment localization research.

To address these challenges, we propose the Cross-modal Semantic Perception Network (SEEN), which simultaneously considers intra-modal semantic modeling and inter-modal semantic collaboration for video moment localization. Specifically, we design a dual-path feature encoding network for multimodal representation and subsequently develop a semantic collaboration network (SCN) for modeling the semantic correlation across modalities. Considering the orthogonality between semantic consistency and complementarity, we integrate inter-modal (frame-to-word and sentence-to-moment) semantic correlation cues and temporal-spatial boundary information to achieve comprehensive horizontal multi-granularity semantic alignment. In parallel, a hierarchical semantic supplement module is introduced to iteratively perform top-down semantic complementarity modeling, thereby enhancing semantic reasoning capabilities. Ultimately, the proposed approach facilitates the effective localization of target moments through comprehensive semantic collaboration.

5.2 Related Work

Video moment localization, which aims at localizing a desired moment within an untrimmed video using a given language query [15], was initially introduced by Hendricks et al. [1] and Gao et al. [5]. These pioneering works addressed the task from two distinct perspectives: **S1**: Shared representation space-based similarity calculation and **S2**: Multi-representation fusion-based alignment prediction. To further enhance localization performance, numerous advanced solutions have been proposed based on these approaches. In what follows, we elaborate on them sequentially.

S1 and Its Representative Evolution The model proposed in [1] directly merged the local moment with the entire video encoding to serve as the context-enhanced visual representation. This approach, however, could introduce noise and reduce discriminative capability. Inspired by this, subsequent researchers proposed several improved models [7, 9, 10, 18, 19, 25, 28, 29, 35]. For instance, Hendricks et al. [7] developed an enhanced moment localization model that treats moment contextual information as an adjustable variable, performing adaptive semantic contextual modeling conditioned on specific video-query semantic similarity. To further improve inter-modal semantic alignment, Hu et al. [10] introduced a cross-modal semantic alignment network model that uses coarse-grained semantic pruning and fine-grained semantic alignment to effectively locate target moments in a coarse-to-fine manner. Meanwhile, Sun et al. [25] proposed a text-query-based video segment retrieval model that conducts common-space video-query semantic similarity evaluation and query-guided semantic alignment to improve video moment localization performance via cross-modal collaborative ranking.

Although the aforementioned approaches have achieved promising performance, they often overlook localization efficiency. To address this issue, Hu et al. [9] proposed a novel deep cross-modal hashing network model for moment localization. This model projects inhomogeneous representations into a shared Hamming space for hash code learning, enabling more efficient retrieval of target moments using Hamming distance calculation instead of traditional similarity metrics. Additionally, Gao et al. [6] presented a knowledge distillation-based moment localization model, which transfers more discriminative inter-modal semantic similarity into the common representation space, thereby achieving more efficient target moment localization.

S2 and Its Representative Evolution The network proposed in [5] concatenates moment representations with query embeddings to jointly align moment-query pairs and regress moment boundaries. Given that simple multi-modal representation concatenation may not fully capture important cross-modal interactions, subsequent researchers have turned to attention mechanisms for effective inter-modal interaction modeling [12, 13, 16, 17, 34, 37, 39].

Specifically, Liu et al. [16, 17] presented two attention-based video moment localization models that utilize visual-aware and query-aware attention mechanisms to capture useful contextual information and interactions for effective semantic fusion, thereby achieving inter-modal semantic alignment and necessary boundary adjustments to localize target moments. Additionally, Yuan et al. [34] introduced a co-attention-based temporal coordinate regression model, employing a co-attention module to capture cross-modal semantic interactions and using attention-guided regression to generate target moment coordinates. Recognizing the limitations of these attention-based models in fully perceiving inter-modal semantic correlations, Lin et al. [13, 40] developed the cross-modal interaction network, which utilizes multi-stage cross-modal interaction and query reconstruction modules to predict moment-query matching scores and adaptively adjust moment boundaries.

To further emphasize the leading role of the language query, Xu et al. [33] proposed a query-guided segment proposal network, which generates query-guided moment candidates and employs multi-task collaborative learning for query-to-moment localization. Subsequently, for a more comprehensive temporal-spatial perception, Zhang et al. [38] designed the 2D Temporal Adjacent Network (2D-TAN), which projects moment representations into a 2D temporal adjacent map to achieve semantic fusion-based moment localization. Although this model incorporates temporal adjacent mapping for moment relationship modeling, it overlooks the importance of moment boundary information. In contrast to previous models, Li et al. [12] proposed a contextual pyramid network (CPNet) to iteratively calibrate temporal boundaries for queried moments using proposal-free regression. Zhang et al. [39] presented an adversarial bi-directional interaction network that captures long-range semantic dependencies and calculates matching scores through adversarial training. Moreover, Zhang et al. [37] introduced a video span localization network that approaches the CML task from a span-based question-answering perspective.

In summary, although substantial progress has been made in video moment localization through the development of these models, they primarily focus on inter-modal semantic consistency while neglecting the necessity of inter-modal semantic complementarity. Consequently, effective modeling of semantic collaboration remains a worthy area of further exploration, which could continue to enhance localization performance.

5.3 Methodology

In this section, we begin by formulating the task of cross-modal moment localization (Sect. 5.3.1). Subsequently, we present our proposed model, SEEN. As illustrated in Fig. 5.2, SEEN primarily consists of two key components: the dual-path representation module (Sect. 5.3.2) and the semantic collaborative learning module (Sect. 5.3.3).

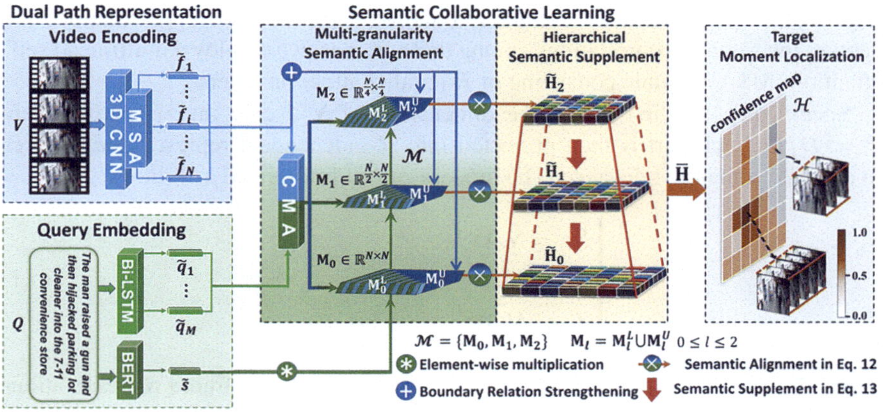

Fig. 5.2 Overview of the framework of the proposed SEEN model

5.3.1 Problem Formulation

Consider a given video and a query sentence, represented as a sequence of video frames and words, respectively. We denote these as $\mathcal{V} = \{f_i\}_{i=1}^{I}$ and $Q = \{w_j\}_{j=1}^{J}$, where I and J indicate the total number of frames and words, respectively. Given these multimedia data, video moment localization aims to locate the target moment θ^* within $\mathcal{V}$ that is semantically aligned with the query Q. Without loss of generality, we assume that the temporal span of the target moment θ^* is $[\tau_s^*, \tau_e^*]$.[1] To address the video moment localization task, we first generate a set of candidate moment representations Θ based on $\mathcal{V}$. We then perform cross-modal semantic perception to identify the optimal moment θ' from the candidate set Θ, ensuring that the temporal spans of θ^* and θ' are identical, i.e., $[\tau_s^*, \tau_e^*] = [\tau_s', \tau_e']$.

5.3.2 Dual Path Representation

As depicted in Fig. 5.2, the dual path representation module comprises two independent encoders, i.e., video encoding and query embedding, that separately perform semantic representation for the given video $\mathcal{V}$ and query Q.

5.3.2.1 Video Encoding

As illustrated in Fig. 5.2, given the untrimmed video $\mathcal{V}$, we first extract its frame features using a 3D CNN extractor [14, 27]. This yields the representation $\mathbf{V}$ =

[1] In this paper, the symbols τ_s^* and τ_e^* represent the start and end timestamps of moment θ^*, respectively.

$[\widehat{\mathbf{f}}_1, \ldots, \widehat{\mathbf{f}}_i, \ldots, \widehat{\mathbf{f}}_N] \in \mathbb{R}^{N \times D}$, which contains N frame-level representations. To capture the semantic correlations among these frames, we employ a multi-head self-attention (MSA) module consisting of R parallel attention layers.

Specifically, we first divide the dimensions of $\mathbf{V}$ evenly into R parts, where $d = D/R$. Each part is then projected into R independent representation spaces to evaluate semantic relevance. The process can be formulated as follows:

$$\begin{cases} \mathbf{X}_r = \mathbf{V}_r \mathbf{W}_r^X, \\ \mathbf{Y}_r = \mathbf{V}_r \mathbf{W}_r^Y, \\ \mathbf{Z}_r = \mathbf{V}_r \mathbf{W}_r^Z, (1 \leq r \leq R), \end{cases} \tag{5.1}$$

where $\mathbf{X}_r \in \mathbb{R}^{N \times d}$, $\mathbf{Y}_r \in \mathbb{R}^{N \times d}$, and $\mathbf{Z}_r \in \mathbb{R}^{N \times d}$ are the frame representations projected into the r-th representation space. The matrices $\mathbf{W}_r^X \in \mathbb{R}^{D \times d}$, $\mathbf{W}_r^Y \in \mathbb{R}^{D \times d}$, and $\mathbf{W}_r^Z \in \mathbb{R}^{D \times d}$ are learnable projection matrices.

Subsequently, we utilize scaled dot-product attention to evaluate the semantic correlations, which is given by:

$$\mathbf{A}_r = Softmax(\frac{\mathbf{X}_r \mathbf{Y}_r^{\top}}{\sqrt{d}})\mathbf{Z}_r, \tag{5.2}$$

where the $Softmax(\cdot)$ operation is applied to each row, and $\mathbf{A}_r \in \mathbb{R}^{N \times d}$ represents the output of the r-th attention layer. Finally, we concatenate the outputs from all R layers to derive the context-enhanced frame-level representations $\widetilde{\mathbf{V}} \in \mathbb{R}^{N \times D}$, as follows:

$$\widetilde{\mathbf{V}} = Concat(\mathbf{A}_1, \ldots, \mathbf{A}_R) + \mathbf{V}, = \{\widetilde{\mathbf{f}}_1, \ldots, \widetilde{\mathbf{f}}_i, \ldots, \widetilde{\mathbf{f}}_N\}. \tag{5.3}$$

5.3.2.2 Query Embedding

The query embedding network is designed to derive both word-level and sentence-level representations.

Word-level Representation To effectively capture context-enhanced semantic information, we initially utilize the GloVe word2vec model [3, 26] to generate embedding vectors for each word. These word embeddings are then fed sequentially into a Bi-LSTM network, which extracts the information from the current word w_j and its context, as formulated below:

$$\begin{cases} \ddot{\mathbf{w}}_j = embedding(w_j), \\ \mathbf{h}_j^{fw} = LSTM^{fw}(\ddot{\mathbf{w}}_t, \mathbf{h}_{t-1}^{fw}), \\ \mathbf{h}_j^{bw} = LSTM^{bw}(\ddot{\mathbf{w}}_t, \mathbf{h}_{t-1}^{bw}), \\ \widehat{\mathbf{q}}_j = [\mathbf{h}_j^{fw}, \mathbf{h}_j^{bw}], \end{cases} \tag{5.4}$$

where $\ddot{\mathbf{w}}_j$ represents the word embedding of w_j, while $\mathbf{h}_j^{fw}$ and $\mathbf{h}_j^{bw}$ denote the forward and backward hidden states of w_j, respectively. We concatenate $\mathbf{h}_j^{fw}$ and $\mathbf{h}_j^{bw}$ to obtain the representation $\widehat{\mathbf{q}}_j$ for w_j. Using this approach, we derive J word-level representations, denoted as $\widehat{\mathbf{Q}} = \{\widehat{\mathbf{q}}_1, \ldots, \widehat{\mathbf{q}}_j, \ldots, \widehat{\mathbf{q}}_J\}$.

To enhance the generalization of these representations, we employ a one-layer fully connected network to obtain the final word-level representation $\widetilde{\mathbf{Q}} \in \mathbb{R}^{J \times D}$, expressed as follows:

$$\widetilde{\mathbf{Q}} = \delta_q(\mathbf{W}_q \widehat{\mathbf{Q}} + b_q), \tag{5.5}$$

where δ_q denotes the ReLU activation function [32], and $\mathbf{W}_q$ and b_q are the weight matrix and bias vector, respectively.

Sentence-level Representation Considering the superior representation capability of BERT for capturing holistic semantic information compared to Bi-LSTM and its derivatives [4], we employ the BERT model [20] to extract the holistic query semantic representation $\bar{\mathbf{s}}$ from the given query. We then use a multi-layer perceptron (MLP) to derive the sentence-level semantic representation $\widetilde{\mathbf{s}}$, formulated as:

$$\begin{cases} \widehat{\mathbf{s}}^1 = \delta_q^1(\mathbf{W}_q^1 \, \bar{\mathbf{s}} + b_q^1), \\ \quad \vdots \\ \widehat{\mathbf{s}}^e = \delta_q^e(\mathbf{W}_q^e \, \widehat{\mathbf{s}}^{e-1} + b_q^e), \\ \quad \vdots \\ \widehat{\mathbf{s}}^E = \delta_q^E(\mathbf{W}_q^E \, \widehat{\mathbf{s}}^{E-1} + b_q^E), \end{cases} \tag{5.6}$$

where $\mathbf{W}_q^e$, b_q^e, $\widehat{\mathbf{s}}^e$, and δ_q^e respectively represent the weight matrix, bias vector, output vector, and ReLU activation function of the e-th hidden layer. The final sentence-level semantic representation, denoted as $\widetilde{\mathbf{s}} = \widehat{\mathbf{s}}^E \in \mathbb{R}^D$, captures the comprehensive semantics of the query, where D is the dimensionality of the representation.

5.3.3 Semantic Collaborative Learning

With the context-enhanced representations of frames $\widetilde{\mathbf{V}} \in \mathbb{R}^{N \times D}$, words $\widetilde{\mathbf{Q}} \in \mathbb{R}^{J \times D}$, and the sentence $\widetilde{\mathbf{s}} \in \mathbb{R}^D$, we introduce SCN to explore semantic relevance between the two modalities. More specifically, SCN performs multi-granularity semantic alignment and hierarchical semantic supplementation to achieve effective inter-modal semantic collaborative learning.

Multi-granularity Semantic Alignment Given that a video typically conveys richer information compared to a query sentence, we employ a cross-modal multi-head attention (CMA) module to capture the key word clues from the query for each video frame. Similar to the MSA mechanism, we divide the feature dimensions D of $\widetilde{\mathbf{V}}$ and $\widetilde{\mathbf{Q}}$ into T equal parts, where $\overline{d} = D/T$. We then project these segments into T joint representation spaces to capture crucial frame-to-word semantic correlations while suppressing non-essential ones. Specifically, for the t-th head of the CMA, the attention calculation is formulated as follows:

$$\widetilde{\mathbf{A}}_t = Softmax(\frac{\widetilde{\mathbf{V}}_t \widetilde{\mathbf{Q}}_t^\top}{\sqrt{\overline{d}}})\widetilde{\mathbf{Q}}_t, \tag{5.7}$$

where $Softmax(\cdot)$ is applied to each row, $\widetilde{\mathbf{V}}_t \in \mathbb{R}^{N\times\overline{d}}$ and $\widetilde{\mathbf{Q}}_t \in \mathbb{R}^{J\times\overline{d}}$ are the inputs, and $\widetilde{\mathbf{A}}_t \in \mathbb{R}^{N\times\overline{d}}$ is the output of the t-th attention head. The essential frame-to-word semantic correlations from the T attention heads are concatenated, and the enhanced representation $\overline{\mathbf{V}} \in \mathbb{R}^{N\times D}$ is obtained as:

$$\overline{\mathbf{V}} = Concat(\widetilde{\mathbf{A}}_1, \ldots, \widetilde{\mathbf{A}}_T) + \widetilde{\mathbf{V}}, = \{\bar{\mathbf{f}}_1, \ldots, \bar{\mathbf{f}}_i, \ldots, \bar{\mathbf{f}}_N\}. \tag{5.8}$$

Inspired by the 2D temporal map structure [39], we propose a competitive multi-layer temporal-spatial matrix $\mathcal{M}$, which comprises a series of temporal-spatial feature maps. Unlike single-scale 2D temporal maps [39] that only contain semantic information in the upper triangular part, our multi-scale temporal-spatial feature maps simultaneously reflect inter-modal semantic correlations and boundary position relations. In other words, $\mathcal{M}$ facilitates multi-granularity moment-to-sentence semantic consistency modeling, as well as temporal-spatial boundary perception enhancement. As shown in Fig. 5.2, $\mathcal{M}$ consists of three layers of temporal-spatial feature maps, denoted as $\mathcal{M} = \{\mathbf{M}_l\}_{0\leq l\leq 2}^{L}$. The dimensions of the three feature maps are $\mathbf{M}_l \in \mathbb{R}^{\frac{N}{2^l}\times\frac{N}{2^l}}$ $(0 \leq l \leq 2)$, with different size scales $\frac{N}{2^l} \times \frac{N}{2^l}$ representing different levels of semantic granularity [10]. Each $\mathbf{M}_l$ consists of upper and lower triangular parts divided by the diagonal line, denoted as $\mathbf{M}_l = \mathbf{M}_l^U \bigcup \mathbf{M}_l^L$. The horizontal and vertical axes of each triangular part represent the starting and ending frame indices, respectively.

We generate multi-granularity moment representation sets and project them into the lower triangular matrices, where $\mathbf{\Theta}_l$ represents the specific moment representation set for $\mathbf{M}_l^L$. Each moment representation in $\mathbf{\Theta}_l$ is denoted by $\mathbf{\Theta}_l = \{\boldsymbol{\theta}_{i,j}^l\}$,[2] where $1 \leq i \leq N/2^l$, $1 \leq j \leq (N/2^l - i + 1)$, and 2^l represents the scaling factor in the l-th layer.

We also generate the corresponding moment boundary representation set $\mathbf{\Delta}_l = \{\widehat{\boldsymbol{\theta}}_{i,j}^l\}$, where $1 \leq i < N/2^l$, $1 \leq j < (N/2^l - i + 1)$. This set supplements necessary

[2] i and j represent the indices of the starting and ending frames, respectively.

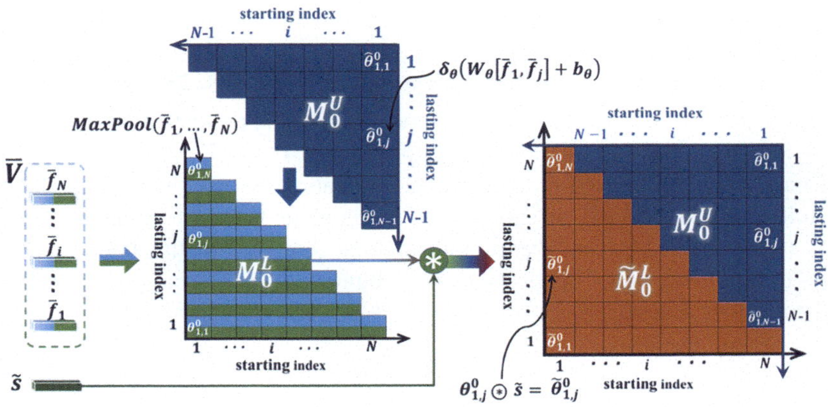

Fig. 5.3 The construction of $\widetilde{\mathbf{M}}_0$, which is projected as a 2D figure to illustrate inter-modal semantic collaboration

boundary information into each upper triangular part $\mathbf{M}_l^U$, further reinforcing temporal-spatial position relations.

To further illustrate this process, we consider the construction of $\mathbf{M}_0$ and $\widetilde{\mathbf{M}}_0$ as an example. As depicted in Fig. 5.3, the representation $\boldsymbol{\theta}_{1,N}^0$ is obtained by max pooling the first N features of $\overline{\mathbf{V}}$, i.e., $\boldsymbol{\theta}_{1,N}^0 = maxpool(\bar{\mathbf{f}}_1, \ldots, \bar{\mathbf{f}}_N)$. Generally, each moment representation $\boldsymbol{\theta}_{i,j}^0 \in \boldsymbol{\Theta}_0$ is generated as follows:

$$\boldsymbol{\theta}_{i,j}^0 = maxpool(\bar{\mathbf{f}}_i, \ldots, \bar{\mathbf{f}}_{i+j-1}). \tag{5.9}$$

After generating all moment representations of $\boldsymbol{\Theta}_0$, we project $\boldsymbol{\Theta}_0$ into the corresponding positions of $\mathbf{M}_0^L$, followed by element-wise multiplication ($\circledast$) for sentence-to-moment semantic fusion:

$$\widetilde{\boldsymbol{\theta}}_{i,j}^0 = \boldsymbol{\theta}_{i,j}^0 \circledast \tilde{\mathbf{s}}. \tag{5.10}$$

Through this operation, we build $\widetilde{\mathbf{M}}_0^L$. Next, we supplement boundary information into the upper triangular part $\mathbf{M}_0^U$ to further enhance temporal-spatial position relations. Specifically, the moment boundary representation set $\boldsymbol{\Delta}_0 = \{\widehat{\boldsymbol{\theta}}_{i,j}^0\}$ is generated, where each representation is computed as:

$$\widehat{\boldsymbol{\theta}}_{i,j}^0 = \delta_\theta(\mathbf{W}_\theta \left[\bar{\mathbf{f}}_i, \bar{\mathbf{f}}_{i+j-1}\right] + b_\theta), \tag{5.11}$$

where $\bar{\mathbf{f}}_i$ and $\bar{\mathbf{f}}_{i+j-1}$ are the boundary features of $\widehat{\boldsymbol{\theta}}_{i,j}^0$, [,] denotes concatenation, δ_θ represents the ReLU activation function, and $\mathbf{W}_\theta$ and b_θ are the weight matrix and bias vector, respectively. After generating $\boldsymbol{\Delta}_0$, we project it into $\mathbf{M}_0^U$ according to its coordinate system. Thus, the first layer of the temporal-spatial feature map ($\mathbf{M}_0 = \widetilde{\mathbf{M}}_0^L \bigcup \mathbf{M}_0^U$) is completed.

Next, we apply a downsampling strategy to generate compact semantic representations $\overline{\mathbf{V}}_l \in \mathbb{R}^{\frac{N}{2^l} \times D}$ for $1 \leq l < L$, and repeat similar operations on $\overline{\mathbf{V}}_l$ to construct the remaining temporal-spatial feature maps $\{\widetilde{\mathbf{M}}_l = \widetilde{\mathbf{M}}_0^L \bigcup \mathbf{M}_0^U\}_{l=1}^L$.

Considering the powerful representation capabilities of convolutional layers, we utilize a CNN model to capture the essential inter-modal semantic correlations (frame-to-word, sentence-to-moment) and temporal-spatial boundary relations, thereby achieving effective semantic alignment. The overall processing flow is expressed as follows:

$$\mathbf{H}_l = Conv\left(\widetilde{\mathbf{M}}_l\right), \quad 0 \leq l < L. \tag{5.12}$$

Hierarchical Semantic Supplement After deriving the multi-layer semantic correlations, we leverage the semantic information from the upper layers to guide complementary semantic modeling in the current layer. Specifically, the top-down semantic supplementation process for the l-th layer temporal-spatial feature matrix can be formulated as follows:

$$\widetilde{\mathbf{H}}_{l-1} = \mathbf{H}_{l-1} \circledast \mathbf{U}(\mathbf{H}_l) \quad 0 \leq l < L, \tag{5.13}$$

where $\mathbf{U}$ and $\circledast$ denote the upsampling operation and element-wise multiplication, respectively. For example, we first apply upsampling to resize $\mathbf{H}_2$ to match the dimensions of $\mathbf{H}_1$, followed by element-wise multiplication ($\circledast$) to perform semantic fusion. Ultimately, the final enhanced representations $\widetilde{\mathbf{H}}_0$, denoted as $\overline{\mathbf{H}}$, achieve comprehensive semantic collaboration by incorporating both alignment and supplementation.

Target Moment Localization After obtaining $\overline{\mathbf{H}}$, we feed it into a CNN module to generate the final semantic similarity confidence map $\mathcal{H} \in \mathbb{R}^{N \times N}$. Each element $\hbar_{i,j} \in \mathcal{H}$ represents the semantic alignment score between a moment candidate $\theta_{i,j}$[3] and the given query. To achieve more accurate alignment scores, we propose a cross-modal semantic collaborative learning loss function $\mathcal{L}$.

Specifically, we first rearrange all the moment candidates in $\mathcal{H}$, expressed as:

$$\mathbf{\Theta} = \{\boldsymbol{\vartheta}_1, \cdots, \boldsymbol{\vartheta}_k, \cdots, \boldsymbol{\vartheta}_K\}, \tag{5.14}$$

where $K = N^2$ is the total number of moment candidates. We then use the commonly adopted Intersection over Union (IoU) metric to assess the alignment between each moment candidate in $\mathbf{\Theta}$ and the given query. More specifically, the

[3] The moment candidate starting at the i-th frame and lasting for j frames.

semantic alignment score Υ_k for the k-th moment and the given query is computed as follows:

$$\Upsilon_k = \frac{\min\left(\tau_e^k, \tau_e^*\right) - \max\left(\tau_s^k, \tau_s^*\right)}{\max\left(\tau_e^k, \tau_e^*\right) - \min\left(\tau_s^k, \tau_s^*\right)}, \tag{5.15}$$

where τ_s^k and τ_e^k indicate the start and end timestamps of the k-th moment candidate, and τ_s^* and τ_e^* denote the ground truth start and end timestamps of the target moment that is semantically consistent with the given query.

Consequently, the semantic alignment loss function $\mathcal{L}$, based on binary cross-entropy, is defined as:

$$\mathcal{L} = \frac{1}{K}\sum_{k=1}^{K}\left[\Upsilon_k \log \tilde{a}_k + (1 - \Upsilon_k)\log(1 - \tilde{a}_k)\right], \tag{5.16}$$

where $\tilde{a}_k$ is the predicted semantic alignment score of the k-th moment candidate.

5.4 Experiments

In this section, we begin by describing the experimental settings. Following this, we conduct comprehensive comparative experiments, extensive ablation studies, and detailed hyperparameter analyses.

5.4.1 Experimental Settings

5.4.1.1 Datasets

The experiments are conducted on three benchmark datasets: ActivityNet Captions, Charades-STA, and TACoS, summarized in Table 5.1.

Table 5.1 Data statistics for the ActivityNet captions, Charades-STA, and TACoS datasets. $\mathcal{N}_V$ denotes the number of videos, $\mathcal{N}_Q$ is the number of queries, $\mathcal{T}_V$ represents the average duration (in seconds) of a video, $\mathcal{T}_M$ denotes the average duration (in seconds) of a moment, and $\mathcal{L}_Q$ is the average length (in words) of a query

Dataset	Video source	$\mathcal{N}_V$	$\mathcal{N}_Q$	$\mathcal{T}_V$	$\mathcal{T}_M$	$\mathcal{L}_Q$
ActivityNet Captions	YouTube	19,811	71,953	117.91	37.81	13.03
Charades-STA	Homes	6670	17,124	30.57	8.00	8.54
TACoS	Lab Kitchen	125	17,837	320.52	29.64	6.23

Table 5.2 Summary of the experimental environment

Name	Configuration
Central processing unit	Intel(R) Xeon(R) Gold 6240C CPU @ 2.60GHz
Graphic processing unit	NVIDIA RTX 2080 Ti × 2
Hard disk	SAMSUNG 970 PRO NVMe M.2 2048 GB
Memory	SAMSUNG DDR4 3200MHz RECC 512 GB
Operation system	Ubuntu 16.04.6
Programming language	python 3.7
Machine learning framework	PyTorch 1.8.1

ActivityNet Captions [21] It consists of 19,811 untrimmed videos and 71,957 natural language queries, covering a wide range of human activities such as eating, drinking, and socializing. The average video and moment durations are 117.74 seconds and 37.14 seconds, respectively, with the average query length being 13 words. We follow the mainstream validation strategy [13], using val_1 for validation and val_2 for testing. This results in 37,417 training, 17,505 validation, and 17,031 testing "moment-query" pairs.

Charades-STA [5] It is built on the Charades dataset [24], which includes 10,000 videos across 157 action categories for temporal action localization. The Charades-STA dataset [5] contains 16,128 "moment-query" pairs, generated by segmenting video-level descriptions into sub-sentences and matching them to corresponding video moments. The average video duration is 30 seconds. We follow the standard split [5], resulting in 12,408 training and 3,720 testing "moment-query" pairs.

TACoS [22] It is derived from the MPII Cooking Composite Activities dataset [23], containing 127 untrimmed cooking videos and 18,818 natural language queries. The average video duration is 296.21 seconds, with target moments averaging 6.10 seconds. The dataset is more challenging due to the short durations and large overlaps between moments in long videos. We follow the standard split [5], forming 10,146 training, 4589 validation, and 4,083 testing "moment-query" pairs.

5.4.1.2 Implementation Details

For the extraction of visual features, we employed the pre-trained C3D network [27] on both the ActivityNet-Captions and TACoS datasets. Additionally, we used three distinct types of visual features, i.e., C3D [27], VGG [36], and I3D [2], for the Charades-STA dataset. Following the methodology outlined in previous work [39], we performed video encoding to generate frame-level representations for each dataset. Specifically, we obtained the following sets of representations: $\mathbf{V}_A \in \mathbb{R}^{64\times512}$ from ActivityNet-Captions, $\mathbf{V}_C \in \mathbb{R}^{64\times512}$ from Charades-STA, and $\mathbf{V}_T \in \mathbb{R}^{128\times512}$ from TACoS. Subsequently, we generated temporal-spatial matrices for each dataset based on these representations. The number of layers (L) for the matrices is set to 2 for ActivityNet-Captions, 2 for Charades-STA, and 3 for TACoS, respectively. For query embeddings, we extracted 512-dimensional word and query features for each dataset. To generate the final confidence maps, we utilized three distinct CNN modules. These modules consist of 4, 8, and 8 convolution layers with kernel sizes of 9, 5, and 5, respectively, for ActivityNet-Captions, Charades-STA, and TACoS. The stride and dilation factors for each convolutional layer are uniformly set to 1 across all three datasets.

During the training phase, we set the batch size to 8, 32, and 32 for the TACoS, Charades-STA, and ActivityNet-Captions datasets, respectively. We optimized the model parameters using the Adam optimizer [11], with a learning rate of 0.0001 for all three datasets. The maximum number of training epochs is set to 600, with early termination employed as necessary to prevent overfitting.

A summary of the main experimental environment is provided in Table 5.2.

5.4.1.3 Evaluation Metrics

We employed the "R@n, IoU=m" metric [8] to evaluate the performance of moment localization. Specifically, for a given query Q_j, the metric $r(n, m, Q_j)$ measures the percentage of the top-n results that achieve an IoU greater than m. To assess the overall performance of the model, we use $R(n, m)$ to represent the average performance of "$R@n$, IoU=m" across all queries. This is computed as:

$$R(n, m) = \frac{1}{N_Q} \sum_{j=1}^{N_Q} r(n, m, Q_j), \tag{5.17}$$

where N_Q denotes the total number of queries, $r(n, m, Q_j)$ represents the recall [30] for the j-th query Q_j, and $R(n, m)$ is the averaged performance over all queries.

5.4.1.4 Baselines

In addition to our proposed model, we compared its performance against the following state-of-the-art baselines:

- **MCN** [1]: This model integrates both local and global video features as visual inputs, projecting both visual and query features into a shared real-valued vector embedding space for moment localization.
- **CTRL** [5]: A cross-modal temporal regression localizer, CTRL evaluates alignment scores between moment candidates and queries, simultaneously adjusting the start and end boundaries of these candidates for localization.
- **ACRN** [16]: This model utilizes a query-aware attention mechanism to enhance visual contextual semantic perception, facilitating effective cross-modal fusion for target moment localization.

- **QSPN** [33]: A query-guided segment proposal network, QSPN first generates moment candidates through a query-guided proposal strategy, then refines the localization using early fusion-based retrieval and multi-task collaborative learning.
- **ABLR** [34]: A co-attention-based temporal coordinates regression model, ABLR uses a co-attention module to capture cross-modal semantic interactions, followed by attention-guided regression for generating the localization coordinates of the target moment.
- **CMIN** [13]: CMIN utilizes multi-head self-attention and syntactic graph convolution networks for separate visual and textual information encoding. It then applies multi-stage cross-modal interaction and query reconstruction modules for moment-query matching score prediction and adaptive adjustment of moment boundaries.
- **2D-TAN** [38]: This model projects the feature representations of the query and moment candidates into a 2D temporal adjacent network for cross-modal feature fusion, and predicts semantic matching scores for moment-query pairs to identify high-scoring moments for localization.
- **CMHN** [9]: This model projects multi-modal representations into a shared Hamming space for hash code learning and leverages efficient Hamming distance calculations for moment localization.
- **CLEAR** [10]: This model adopts a coarse-to-fine approach, utilizing coarse-grained semantic pruning followed by fine-grained semantic alignment to effectively locate target moments.
- **ABIN** [39]: This model introduces a bidirectional guided attention mechanism for interaction modeling and incorporates an auxiliary discriminator network to calibrate the localization results.
- **VSLNet** [37]: This model uses a context-query attention mechanism to capture cross-modal interactions and applies a conditioned span predictor along with a query-guided highlighting strategy for accurate span localization.
- **FVMR** [6]: This model employs a hierarchical semantic attentive mechanism to derive discriminative representations, followed by a matching-score distillation strategy for improved moment retrieval within a shared representation space.

In addition, we provide the following six supplementary instructions regarding the aforementioned baselines: (1) The localization results of the baselines are directly quoted from their original publications. (2) Since ABLR uses co-attention guided regression to directly predict the target moment coordinates based on the maximum attentive weights/features, it only produces R@1 results for a given query. (3) The 2D-TAN model contains two derivations [38], and we report the average $R(n, m)$ across both as its overall performance. (4) As both ABIN and VSLNet localize the target moment by selecting the coordinates with the highest probability, they only return R@1 results for each given query. (5) Due to the absence of comparison experiments for some baselines (MCN, ABLR, ABIN, CMHN, and CLEAR) on the Charades-STA dataset, we reproduced their experimental results on Charades-STA to facilitate the subsequent performance comparison. (6) Given that different baselines utilize varying video features for moment localization on the

Charades-STA dataset, we employ the corresponding features (i.e., C3D, VGG, and I3D) for a fair performance comparison.

5.4.2 Performance Comparison

In this subsection, we presented a comparative analysis of our model's performance on three benchmark datasets. The localization results are summarized in Tables 5.3, 5.4, and 5.5 (with the best results in **bold** and the second-best in underlined). Based on these results, we make the following four key observations:

Table 5.3 Performance comparison on ActivityNet Captions (p-value*: p-value over $R(1, 0.5)$)

Method	R@1 IoU=0.3	R@1 IoU=0.5	R@1 IoU=0.7	R@5 IoU=0.3	R@5 IoU=0.5	R@5 IoU=0.7	p-value*
MCN [1]	39.35%	21.36%	6.43%	68.12%	53.23%	29.70%	2.07E-31
CTRL [5]	47.43%	29.01%	10.34%	75.32%	59.17%	37.54%	2.18E-28
ACRN [16]	49.70%	31.67%	11.25%	76.50%	60.34%	38.57%	5.49E-27
QSPN [33]	52.13%	33.26%	13.43%	77.72%	62.39%	40.78%	5.18E-26
ABLR [34]	55.67%	36.79%	15.71%	–	–	–	2.78E-23
CMIN [13]	**64.41%**	44.62%	24.48%	82.39%	69.66%	52.96%	1.31E-04
2D-TAN [38]	59.10%	44.28%	26.96%	85.59%	76.89%	62.11%	3.79E-07
CMHN [9]	62.49%	43.47%	24.02%	85.37%	73.42%	53.16%	2.95E-11
CLEAR [10]	59.96%	45.33 %	28.05 %	85.83%	77.26%	62.13 %	1.02E-06
ABIN [39]	63.19 %	44.02%	24.23%	–	–	–	1.29E-08
VSLNet [37]	63.16%	43.22%	26.16%	–	–	–	3.11E-12
FVMR [6]	60.63%	45.00%	26.85%	86.11 %	77.42 %	61.04%	5.61E-03
SEEN	61.15%	**45.56%**	**28.10%**	**86.15%**	**77.70%**	**63.35%**	

Table 5.4 Performance comparison on TACoS (p-value*: p-value over $R(1, 0.5)$)

Method	R@1 IoU=0.1	R@1 IoU=0.3	R@1 IoU=0.5	R@5 IoU=0.1	R@5 IoU=0.3	R@5 IoU=0.5	p-value*
MCN [1]	3.11%	1.64%	1.25%	3.35%	2.03	1.25%	2.56E-44
CTRL [5]	24.32%	18.32%	13.30%	48.73%	36.69%	25.42%	4.86E-39
ACRN [16]	24.22%	19.52%	14.62%	47.42%	34.97%	24.88%	2.35E-38
QSPN [33]	25.31%	20.15%	15.23%	53.21%	36.72%	25.30%	1.56E-37
ABLR [34]	34.70%	19.50%	9.40%	–	–	–	3.86E-41
CMIN [13]	36.88%	27.33%	19.57%	64.93%	43.35%	28.53%	3.74E-33
2D-TAN [38]	47.02%	36.26%	25.26%	72.37%	57.38%	44.63%	9.64E-09
CMHN [9]	40.19%	30.04%	25.58%	65.33%	44.05%	35.23%	1.10E-08
CLEAR [10]	49.11%	42.18 %	30.27 %	73.56%	63.61%	51.76 %	3.22E-04
ABIN [39]	32.77%	23.63%	20.16%	–	–	–	5.63E-33
VSLNet [37]	50.51%	29.61%	24.27%	–	–	–	4.70E-20
FVMR [6]	53.12 %	41.48%	29.12%	78.12 %	64.53 %	50.00%	3.97E-05
SEEN	**55.66%**	**43.26%**	**31.15%**	**78.32%**	**66.33%**	**52.78%**	

Table 5.5 Performance comparison on Charades-STA (p-value*: p-value over $R(1, 0.5)$)

Method	Feature	R@1 IoU=0.5	R@1 IoU=0.7	R@5 IoU=0.5	R@5 IoU=0.7	p-value*
CTRL [5]	C3D	23.63%	8.89%	58.92%	29.52%	2.93E-26
ACRN [16]	C3D	20.26%	7.64%	71.99%	27.79%	1.25E-27
QSPN [33]	C3D	35.60 %	15.80 %	79.40 %	45.40 %	2.08E-17
ABLR [34]	C3D	24.36%	9.01%	–	–	1.97E-10
SEEN	C3D	**39.54%**	**22.88%**	**81.96%**	**52.23%**	
MCN [1]	VGG	17.46%	8.01%	48.22%	26.73%	1.32E-28
2D-TAN [38]	VGG	39.76%	23.28%	79.83%	51.71%	6.03E-08
CMHN [9]	VGG	29.33%	10.46%	70.43%	38.28%	3.30E-23
CLEAR [10]	VGG	38.46%	22.15%	77.67%	53.02 %	7.15E-09
FVMR [6]	VGG	42.36 %	24.14 %	83.97 %	50.15%	4.62E-12
SEEN	VGG	**43.04%**	**25.89%**	**85.40%**	**56.13%**	
VSLNet [37]	I3D	47.31 %	30.19 %	–	–	5.41E-05
SEEN	I3D	**54.11%**	**33.44%**	**88.71%**	**61.96%**	

- MCN and CTRL exhibit the lowest localization performance across all three datasets. This performance limitation can be attributed to their relatively weak intra-modal semantic understanding. Specifically, MCN combines moment features with the entire video semantic representation, which may introduce noise and reduce its discriminative power. In contrast, CTRL selects only limited pre- and post-contextual information, which prevents it from effectively capturing the full contextual semantic dependencies, leading to suboptimal localization performance.
- The performance of attention-based models (ABLR, QSPN, ACRN, and ABIN) is generally superior to that of MCN and CTRL. These models leverage attention mechanisms to highlight critical semantic information, facilitating better cross-modal interaction. However, relying solely on attention mechanisms does not suffice for accurately capturing complex inter-modal semantic correlations. This limitation is evident, for instance, in the substantial performance drop of ABLR on $R(1, 0.5)$ in Table 5.4.
- Models such as 2D-TAN, CMHN, CMIN, CLEAR, VSLNet, and FVMR achieve better performance than the aforementioned baselines. Their success can be attributed to innovations such as temporal adjacency relation modeling, long-range semantic dependencies, and coarse-to-fine cross-modal interaction, all of which contribute to improved localization accuracy.
- Our proposed SEEN model achieves the best performance, surpassing all other state-of-the-art baselines. Compared to the best-performing model on $R(1, 0.5)$, SEEN achieves average performance improvements of 3.81, 0.88, and 0.23% on Charades-STA, TACoS, and ActivityNet-Captions, respectively. These improvements demonstrate the effectiveness of our semantic collaborative learning

approach, which enhances cross-modal semantic alignment and supplementation, leading to significantly improved localization performance.

Furthermore, we conducted significance tests between SEEN and each of the competing models on $R(1, 0.5)$, based on 20 rounds of testing. The p-values for all comparisons across the three datasets are below 0.05, indicating that the performance improvements of SEEN over the competitors are statistically significant.

5.4.3 Ablation Study

To investigate the contributions of various components in our SEEN model, we conducted an ablation study by designing five different variants of the SEEN model as follows:

- **SEEN-A**: In this variant, we removed the MSA-based contextual semantic enhanced representations $\widetilde{\mathbf{V}}$ in Eq. (5.3) and directly utilized the original frame representations $\mathbf{V}$ for moment localization.
- **SEEN-C**: We eliminated the CMA-based frame-to-word semantic consistency modeling in Eq. (5.7) and the resulting enhanced representations in Eq. (5.8). This variant thus relies solely on sentence-to-moment semantic consistency modeling and temporal-spatial boundary strengthening for moment localization.
- **SEEN-M**: In this case, we removed all upper triangular parts $\mathbf{M}_l^U$ $(0 \leq l \leq L)$ from the temporal-spatial matrix $\mathcal{M}$, effectively deleting all temporal-spatial boundary information. Only the multi-layer lower triangular parts $\mathbf{M}_l^L$ $(0 \leq l \leq L)$ were used for moment localization.
- **SEEN-S**: This variant omits the sentence-to-moment semantic representation from Eq. (5.10) and relies solely on frame-to-word semantic consistency modeling and temporal-spatial boundary strengthening for moment localization.
- **SEEN-W**: We replace the original word-level representation strategy (i.e., word2vec combined with Bi-LSTM) in Eq. (5.4) with BERT-based word embedding extraction for enhanced moment localization.

In addition, on the Charades-STA dataset, since SEEN utilizes three different feature types (C3D, VGG, and I3D) for video moment localization, each variant of SEEN and its ablations adopts distinct video features for video moment localization. The averaged performance across these feature sets is used for the ablation analysis.

We conducted performance comparison experiments on ActivityNet Captions, Charades-STA, and TACoS. The corresponding results and ablation visualization examples are provided in Table 5.6 (with the best results in **bold** and second-best results underlined) and Fig. 5.4. Through a joint analysis of these results and visualizations, we observed the following four key points:

- SEEN-M and SEEN-A exhibit the lowest localization performance across all three datasets, indicating that disregarding contextual or boundary information impairs semantic correlation discrimination. As shown in Fig. 5.4, these two

Table 5.6 Ablation studies on three datasets in terms of $R(1, m)$ $m \in \{0.1, 0.3, 0.5, 0.7\}$

Dataset	Method	R@1 IoU=0.1	R@1 IoU=0.3	R@1 IoU=0.5	R@1 IoU=0.7
ActivityNet captions	SEEN-A	77.38%	59.43%	44.88%	27.22%
	SEEN-C	77.40%	59.81%	44.95%	27.98%
	SEEN-M	76.29%	58.70%	44.39%	26.56%
	SEEN-S	78.73%	59.48%	45.22%	26.48%
	SEEN-W	79.05%	60.03%	45.41%	28.02%
	SEEN	**79.23%**	**61.15%**	**45.56%**	**28.10%**
Charades-STA	SEEN-A	67.02%	58.92%	44.80%	26.93%
	SEEN-C	66.91%	59.04%	44.87%	27.03%
	SEEN-M	66.35%	58.12%	43.24%	26.89%
	SEEN-S	67.84%	58.38%	44.31%	26.34%
	SEEN-W	68.13%	59.15%	45.21%	27.10%
	SEEN	**68.31%**	**59.50%**	**45.56%**	**27.40%**
TACOS	SEEN-A	55.01%	42.08%	28.83%	12.65%
	SEEN-C	55.38%	43.06%	29.88%	12.54%
	SEEN-M	54.97%	42.02%	26.08%	12.21%
	SEEN-S	55.52%	42.12%	28.94%	12.76%
	SEEN-W	55.57%	42.82%	30.26%	13.99%
	SEEN	**55.66%**	**43.26%**	**31.15%**	**14.37%**

variants struggle to capture complex cross-modal semantic correlations, leading to significant discrepancies between the predicted moments and the target moments.

- SEEN-C and SEEN-S achieve suboptimal performance, which suggests that neglecting either frame-to-word or sentence-to-moment semantic alignment limits the model's ability to perceive inter-modal semantics effectively. As demonstrated in Fig. 5.4, these variants fail to perform accurate cross-modal semantic consistency modeling, resulting in low-quality moment localization.
- SEEN-W outperforms the previous four variants, indicating that replacing the word-level representation strategy has a minimal impact on query embedding and overall localization performance. This observation highlights the robustness of our model.
- SEEN achieves the best localization performance, validating that our proposed strategies—such as visual/textual semantic contextual enhancement, frame-to-word and sentence-to-moment semantic consistency modeling, and temporal-spatial boundary information strengthening—substantially improve moment localization accuracy.

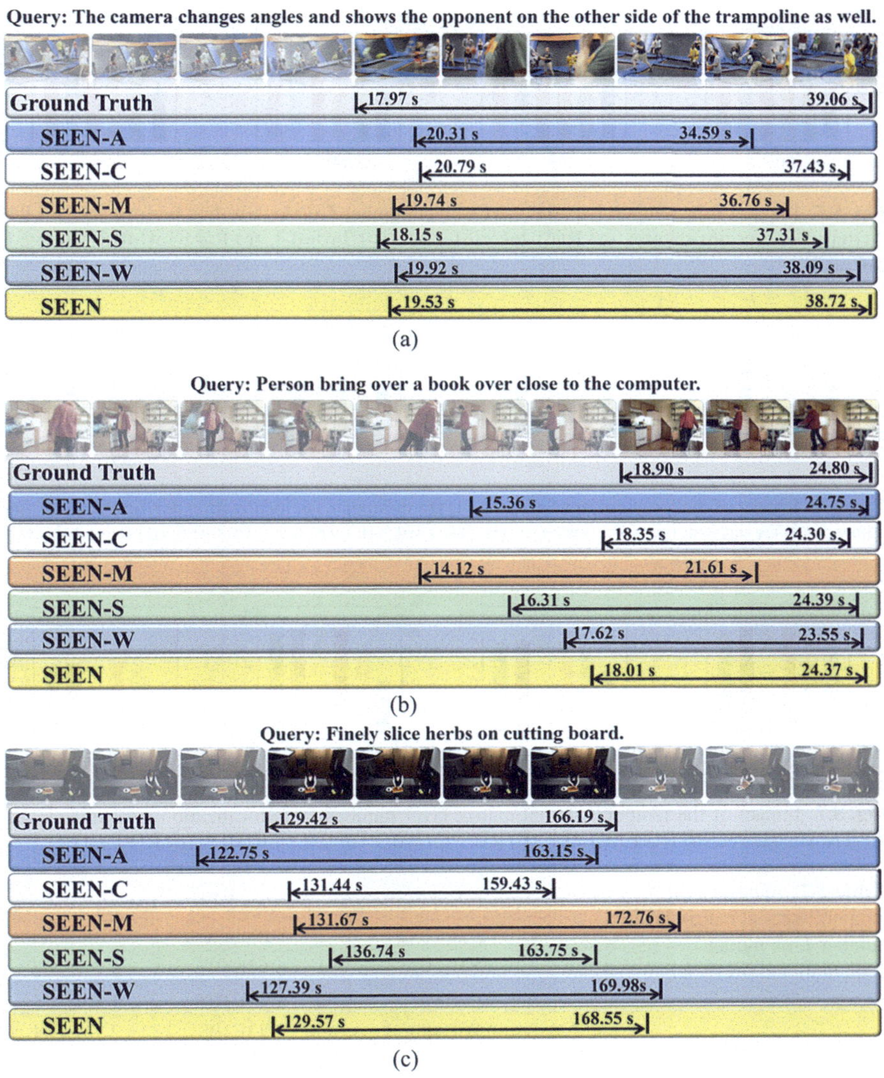

Fig. 5.4 Ablation study results visualized across three datasets. (**a**) The ablation visualization examples on ActivityNet captions. (**b**) The ablation visualization examples on Charades-STA. (**c**) The ablation visualization examples on TACoS

5.4.4 Parameter Analysis

In this subsection, we further investigated the impact of the temporal-spatial matrix layer number, denoted as L, on localization performance across the ActivityNet Captions, Charades-STA, and TACoS datasets. To this end, we conducted local-

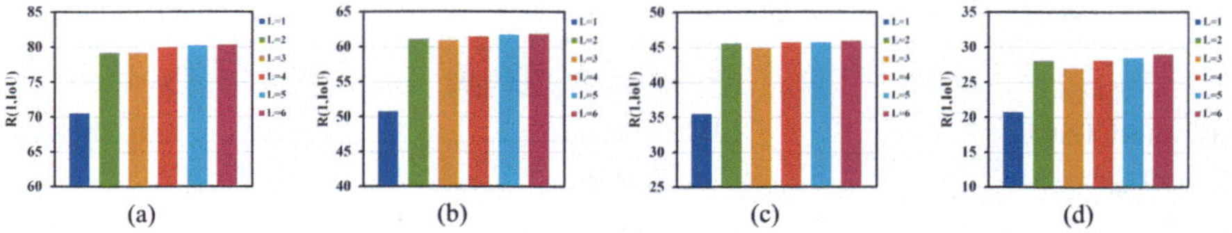

Fig. 5.5 Impact of the temporal-spatial matrix layer number L on localization performance for the ActivityNet captions dataset. (**a**) R@1 IoU=0.1. (**b**) R@1 IoU=0.3. (**c**) R@1 IoU=0.5. (**d**) R@1 IoU=0.7

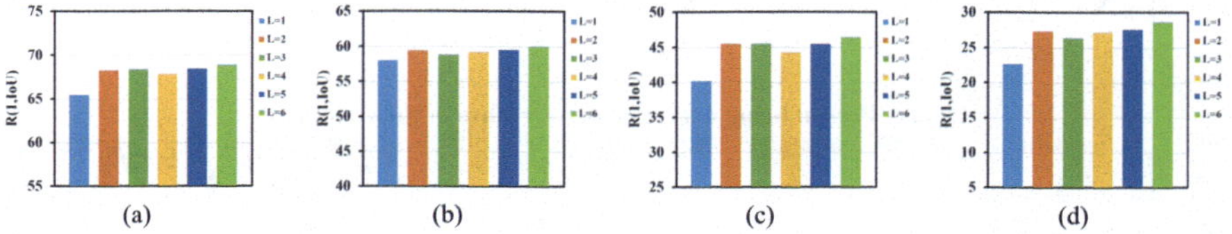

Fig. 5.6 Impact of the temporal-spatial matrix layer number L on localization performance for the Charades-STA dataset. (**a**) R@1 IoU=0.1. (**b**) R@1 IoU=0.3. (**c**) R@1 IoU=0.5. (**d**) R@1 IoU=0.7

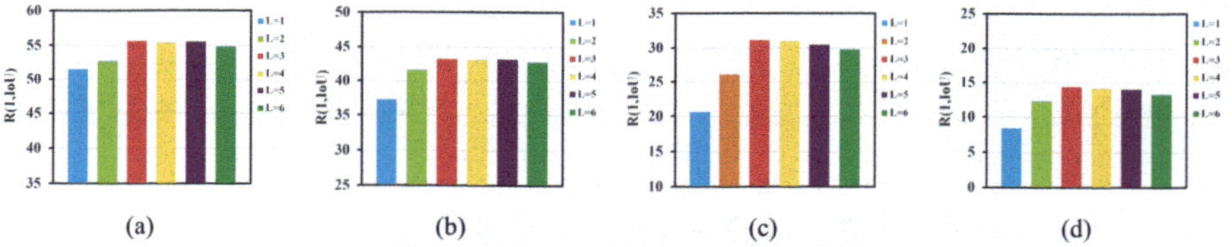

Fig. 5.7 Impact of the temporal-spatial matrix layer number L on localization performance for the TACoS dataset. (**a**) R@1 IoU=0.1. (**b**) R@1 IoU=0.3. (**c**) R@1 IoU=0.5. (**d**) R@1 IoU=0.7

Table 5.7 Impact of the temporal-spatial matrix layer number L on model computational overhead

ActivityNet captions		Charades-STA		TACoS	
L	Parameters	L	Parameters	L	Parameters
1	9.84E+07	1	6.62E+07	1	6.77E+07
2	2.09E+08	2	1.44E+08	2	1.46E+08
3	3.70E+08	3	2.73E+08	3	2.75E+08
4	7.33E+08	4	6.03E+08	4	6.04E+08
5	1.90E+09	5	1.74E+09	5	1.74E+09
6	6.29E+09	6	6.09E+09	6	6.10E+09

ization performance comparison experiments on these benchmark datasets using various evaluation metrics, with L ranging from 1 to 6. The corresponding results for localization performance and model overhead are presented in Figs. 5.5, 5.6, 5.7, and Table 5.7. Based on these results, we observe the following key points:

- As the layer number L increases from 1 to 3, the overall localization performance, evaluated using the four metrics: "R@1 IoU={0.1, 0.3, 0.5, 0.7}", consistently improves across all datasets. Notably, the most significant improvement occurs when L increases from 1 to 2. As analyzed previously, the single-layer ($L = 1$) SEEN model can only perform semantic alignment in a horizontal direction. In contrast, the two-layer ($L = 2$) SEEN model facilitates multi-granularity semantic alignment and hierarchical semantic enhancement, leading to more effective video moment localization. Therefore, the performance boost from $L = 1$ to $L = 2$ highlights the effectiveness of semantic collaborative modeling in two orthogonal directions (horizontal and vertical) and underscores the importance of top-down hierarchical semantic supplementation for video moment localization.

- As L increases further to 6, the overall localization performance fluctuates slightly and experiences slower improvement on the ActivityNet Captions and Charades-STA datasets. In contrast, on the TACoS dataset, the performance remains stable before slightly declining. This suggests that, compared to ActivityNet Captions and Charades-STA, which feature diverse multimedia content, deeper semantic collaborative learning on TACoS (focused on cooking activities) may overfit to trivial semantic details, diminishing the model's ability to capture significant cross-modal correlations, thereby degrading localization performance.
- With the increase in L, the number of model parameters grows substantially. This results in a sharp increase in computational overhead, highlighting the trade-off between deeper semantic learning and the associated computational cost.

To strike a balance between localization performance and computational overhead, we set L to 2, 2, and 3 for ActivityNet Captions, Charades-STA, and TACoS, respectively.

5.4.5 Qualitative Reuslts

To further qualitatively assess the effectiveness of our SEEN model, we selected four baseline models: CMIN, 2D-TAN, FVMR, and VSLNet that demonstrate relatively higher performance. We then compared SEEN with these baselines on the benchmark datasets, presenting three representative moment localization visualization examples.

As shown in the localization results in Fig. 5.8, although the four baseline models successfully localize the corresponding moments in the three examples, their localization performance remains suboptimal. This may be attributed to their reliance on semantic consistency cues for inter-modal alignment, while neglecting the critical aspect of semantic supplementation. In contrast, SEEN achieves superior localization performance. The reasons for this are summarized as follows:

- SEEN effectively leverages multi-granularity semantic correlations and boundary relation strengthening to facilitate robust inter-modal semantic alignment.

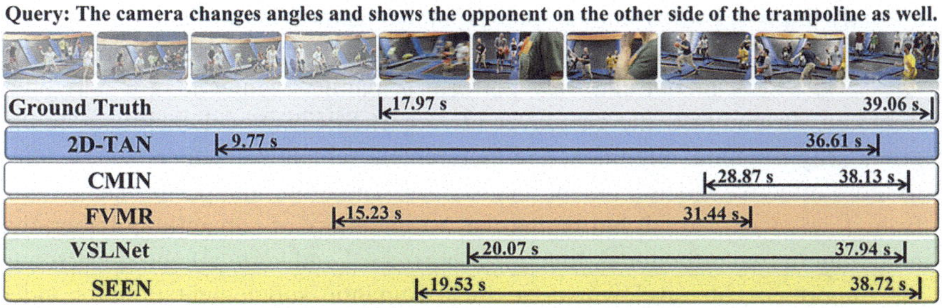

(a) The visualization example on ActivityNet Captions.

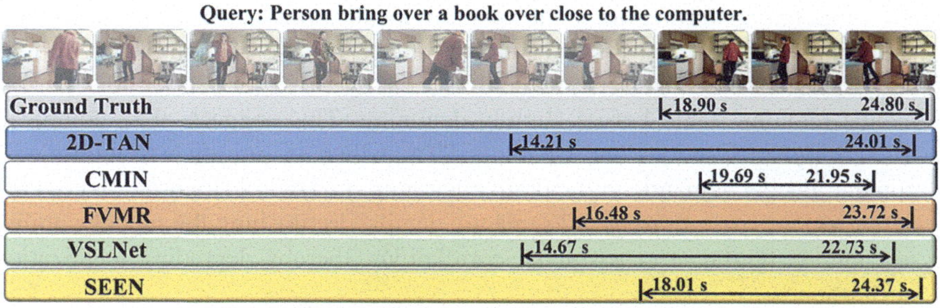

(b) The visualization example on Charades-STA.

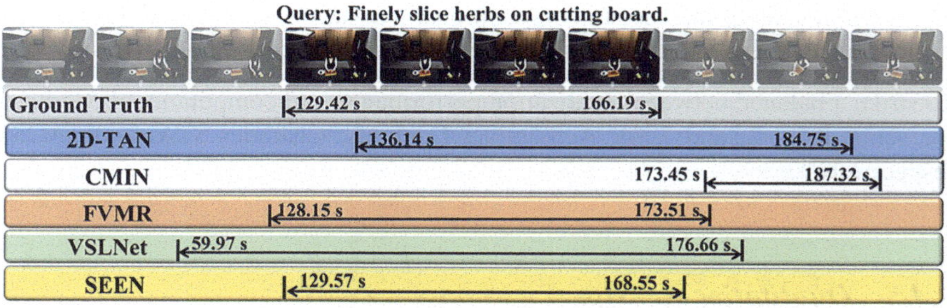

(c) The visualization example on TACoS.

Fig. 5.8 Qualitative comparison results on three benchmark datasets. (**a**) The visualization examples on ActivityNet captions. (**b**) The visualization examples on Charades-STA. (**c**) The visualization examples on TACoS

- SEEN employs hierarchical semantic complementarity modeling to iteratively perform top-down semantic supplementation, resulting in high-quality moment localization.

These qualitative visualization results further validate the effectiveness of our model.

5.5 Summary

In this chapter, we propose a novel cross-modal semantic perception network for video moment localization. First, we introduce a multi-modal representation network that facilitates intra-modal semantic understanding and feature extraction. We then develop a semantic collaboration network that enables comprehensive cross-modal semantic learning, ensuring effective moment localization. Our approach not only captures frame-to-word and sentence-to-moment semantic correlations for inter-modal alignment but also leverages hierarchical semantic complementarity to enrich inter-modal interactions. Extensive experimental comparisons demonstrate the superior performance of our proposed model.

References

1. Anne Hendricks, L., Wang, O., Shechtman, E., Sivic, J., Darrell, T., Russell, B.: Localizing moments in video with natural language. In: Proceedings of the IEEE International Conference on Computer Vision, pp. 5803–5812 (2017)
2. Carreira, J., Zisserman, A.: Quo vadis, action recognition? A new model and the kinetics dataset. In: Proceedings of the IEEE Conference on Computer Vision and Pattern Recognition, pp. 6299–6308 (2017)
3. Chen, X., Song, X., Ren, R., Zhu, L., Cheng, Z., Nie, L.: Fine-grained privacy detection with graph-regularized hierarchical attentive representation learning. ACM Trans. Inf. Syst. **38**, 1–26 (2020)
4. Devlin, J., Chang, M.W., Lee, K., Toutanova, K.: Bert: Pre-training of deep bidirectional transformers for language understanding. In: Proceedings of the Conference of the North American Chapter of the Association for Computational Linguistics: Human Language Technologies, pp. 4171–4186 (2019)
5. Gao, J., Sun, C., Yang, Z., Nevatia, R.: Tall: temporal activity localization via language query. In: Proceedings of the IEEE International Conference on Computer Vision, pp. 5267–5275 (2017)
6. Gao, J., Xu, C.: Fast video moment retrieval. In: Proceedings of the IEEE International Conference on Computer Vision, pp. 1523–1532 (2021)
7. Hendricks, L.A., Wang, O., Shechtman, E., Sivic, J., Darrell, T., Russell, B.: Localizing moments in video with temporal language. In: Proceedings of the International Conference on Empirical Methods in Natural Language Processing, pp. 1380–1390 (2018)
8. Hu, R., Xu, H., Rohrbach, M., Feng, J., Saenko, K., Darrell, T.: Natural language object retrieval. In: Proceedings of the IEEE Conference on Computer Vision and Pattern Recognition, pp. 4555–4564 (2016)
9. Hu, Y., Liu, M., Su, X., Gao, Z., Nie, L.: Video moment localization via deep cross-modal hashing. IEEE Trans. Image Process. **30**, 4667–4677 (2021)
10. Hu, Y., Nie, L., Liu, M., Wang, K., Wang, Y., Hua, X.: Coarse-to-fine semantic alignment for cross-modal moment localization. IEEE Trans. Image Process. **30**, 5933–5943 (2021)
11. Kingma, D.P., Ba, J.: Adam: a method for stochastic optimization. In: Proceedings of the International Conference on Learning Representations, pp. 1–15 (2015)
12. Li, K., Guo, D., Wang, M.: Proposal-free video grounding with contextual pyramid network. In: Proceedings of the American Association for Artificial Intelligence, pp. 1902–1910 (2021)
13. Lin, Z., Zhao, Z., Zhang, Z., Zhang, Z., Cai, D.: Moment retrieval via cross-modal interaction networks with query reconstruction. IEEE Trans. Image Process. **29**, 3750–3762 (2020)

14. Liu, M., Nie, L., Wang, M., Chen, B.: Towards micro-video understanding by joint sequential-sparse modeling. In: Proceedings of the ACM International Conference on Multimedia, pp. 970–978 (2017)
15. Liu, M., Nie, L., Wang, Y., Wang, M., Rui, Y.: A survey on video moment localization. ACM Comput. Surv. **55**, 1–35 (2022)
16. Liu, M., Wang, X., Nie, L., He, X., Chen, B., Chua, T.S.: Attentive moment retrieval in videos. In: Proceedings of the ACM International Conference on Research and Development in Information Retrieval, pp. 15–24 (2018)
17. Liu, M., Wang, X., Nie, L., Tian, Q., Chen, B., Chua, T.: Cross-modal moment localization in videos. In: Proceedings of the ACM International Conference on Multimedia, pp. 843–851 (2018)
18. Mithun, N.C., Paul, S., Roy-Chowdhury, A.K.: Weakly supervised video moment retrieval from text queries. In: Proceedings of the IEEE Conference on Computer Vision and Pattern Recognition, pp. 11592–11601 (2019)
19. Mun, J., Cho, M., Han, B.: Local-global video-text interactions for temporal grounding. In: Proceedings of the IEEE Conference on Computer Vision and Pattern Recognition, pp. 10810–10819 (2020)
20. Qu, L., Liu, M., Cao, D., Nie, L., Tian, Q.: Context-aware multi-view summarization network for image-text matching. In: Proceedings of the ACM International Conference on Multimedia, pp. 1047–1055 (2020)
21. Ranjay, K., Kenji, H., Frederic, R., Li, F.F., Juan, C.N.: Dense-captioning events in videos. In: Proceedings of the IEEE International Conference on Computer Vision, pp. 706–715 (2017)
22. Regneri, M., Rohrbach, M., Wetzel, D., Thater, S., Schiele, B., Pinkal, M.: Grounding action descriptions in videos. Trans. Assoc. Comput. Linguist. **1**, 25–36 (2013)
23. Rohrbach, M., Regneri, M., Andriluka, M., Amin, S., Pinkal, M., Schiele, B.: Script data for attribute-based recognition of composite activities. In: European Conference on Computer Vision, pp. 144–157 (2012)
24. Sigurdsson, G.A., Varol, G., Wang, X., Farhadi, A., Laptev, I., Gupta, A.: Hollywood in homes: crowdsourcing data collection for activity understanding. In: Proceedings of the European Conference on Computer Vision, pp. 510–526 (2016)
25. Sun, X., Long, X., He, D., Wei, S., Lian, Z.: Vsrnet: End-to-end video segment retrieval with text query. Pattern Recognit. **119**, 108027–108036 (2021)
26. Tang, H., Zhu, J., Wang, L., Zheng, Q., Zhang, T.: Multi-level query interaction for temporal language grounding. IEEE Trans. Intell. Transp. Syst. **1**, 1–10 (2021)
27. Tran, D., Bourdev, L., Fergus, R., Torresani, L., Paluri, M.: Learning spatiotemporal features with 3d convolutional networks. In: Proceedings of the IEEE International Conference on Computer Vision, pp. 4489–4497 (2015)
28. Wang, H., Zha, Z.J., Chen, X., Xiong, Z., Luo, J.: Dual path interaction network for video moment localization. In: Proceedings of the ACM International Conference on Multimedia, pp. 4116–4124 (2020)
29. Wang, H., Zha, Z.J., Li, L., Liu, D., Luo, J.: Structured multi-level interaction network for video moment localization via language query. In: Proceedings of the IEEE Conference on Computer Vision and Pattern Recognition, pp. 7026–7035 (2021)
30. Wang, X., Nie, L., Song, X., Zhang, D., Chua, T.S.: Unifying virtual and physical worlds: learning toward local and global consistency. ACM Trans. Inf. Syst. **36**, 1–26 (2017)
31. Wei, Y., Wang, X., Guan, W., Nie, L., Lin, Z., Chen, B.: Neural multimodal cooperative learning toward micro-video understanding. IEEE Trans. Image Process. **29**, 1–14 (2020)
32. Xu, B., Wang, N., Chen, T., Li, M.: Empirical evaluation of rectified activations in convolutional network. In: Proceedings of the International Conference on Machine Learning, pp. 1–5 (2015)
33. Xu, H., He, K., Plummer, B.A., Sigal, L., Sclaroff, S., Saenko, K.: Multilevel language and vision integration for text-to-clip retrieval. In: Proceedings of the American Association for Artificial Intelligence, pp. 9062–9069 (2019)

34. Yuan, Y., Mei, T., Zhu, W.: To find where you talk: temporal sentence localization in video with attention based location regression. In: Proceedings of the American Association for Artificial Intelligence, pp. 9159–9166 (2019)
35. Zeng, R., Xu, H., Huang, W., Chen, P., Tan, M., Gan, C.: Dense regression network for video grounding. In: Proceedings of the IEEE Conference on Computer Vision and Pattern Recognition, pp. 10287–10296 (2020)
36. Zhang, D., Dai, X., Wang, X., Wang, Y.F., Davis, L.S.: Man: moment alignment network for natural language moment retrieval via iterative graph adjustment. In: Proceedings of the IEEE Conference on Computer Vision and Pattern Recognition, pp. 1247–1257 (2019)
37. Zhang, H., Sun, A., Jing, W., Zhou, J.T.: Span-based localizing network for natural language video localization. In: Proceedings of the Association for Computational Linguistics, pp. 6543–6554 (2020)
38. Zhang, S., Peng, H., Fu, J., Luo, J.: Learning 2d temporal adjacent networks for moment localization with natural language. In: Proceedings of the American Association for Artificial Intelligence, pp. 12870–12877 (2020)
39. Zhang, Z., Zhao, Z., Zhang, Z., Lin, Z., Wang, Q., Hong, R.: Temporal textual localization in video via adversarial bi-directional interaction networks. IEEE Trans. Multimedia **23**, 3306–3317 (2020)
40. Zhu, Z., Lin, Z., Zhao, Z., Xiao, Z.: Cross-modal interaction networks for query-based moment retrieval in videos. In: Proceedings of the International Conference on Research and Development in Information Retrieval, pp. 655–664 (2019)

Chapter 6
Weakly-Supervised Video Moment Localization

6.1 Introduction

Video moment localization, which aims to identify specific start and end points within a video in response to a given query, has garnered significant research attention due to its wide range of applications, including video surveillance [2, 3, 8, 17, 18, 23, 26, 34, 37, 39]. Despite the promising progress, most existing methods rely on supervised learning, requiring temporal annotations for each target moment (i.e., start and end points). Manually annotating these temporal boundaries is not only time-consuming but also prone to inconsistencies, as annotators may have different interpretations of when activities begin and end. For example, as illustrated in Fig. 6.1, two human annotators may provide entirely different labels for the same query, such as "The bikers begin the race." This inconsistency introduces bias in the training data, which can negatively affect model performance.

To address these challenges, recent research has shifted toward weakly-supervised models, which require only video-level annotations, as shown in Fig. 6.1. Compared to supervised methods, weakly-supervised video moment localization is more challenging for several reasons:

1. **Video-Query Alignment**: In the weakly-supervised setting, only the video corresponding to the query is provided, without specific moment annotations. However, in real-world scenarios such as autonomous driving or surveillance, untrimmed videos often contain complex scenes with numerous objects and actions. Only a small part of the video may contain relevant information for the query, making video-query alignment a critical and under-explored problem.
2. **Moment-Query Alignment**: Different moments in a video have varying durations and spatio-temporal characteristics [24], making it difficult to identify the relevant moment and estimate its alignment with the query.

M. Liu et al., *Intelligent and Efficient Video Moment Localization*,
https://doi.org/10.1007/978-3-031-87588-5_6

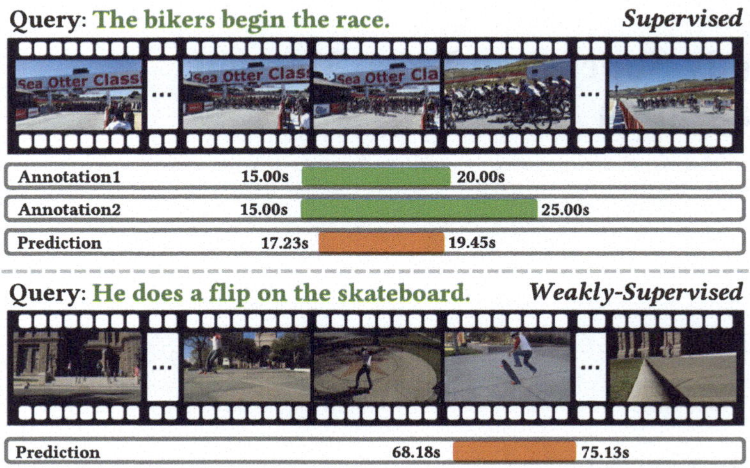

Fig. 6.1 Illustration of supervised and weakly-supervised video moment localization. In the supervised setting, temporal annotations (start and end points) are provided for each moment, whereas in the weakly-supervised setting, only video-level annotations are available, requiring the model to localize the moment from unannotated video content

Although deep learning techniques have shown promise in weakly-supervised video moment retrieval [9, 15, 21, 36], they suffer from several limitations:

1. **Embedding-based Methods**: Many approaches [15, 36] map both moments and queries into a joint embedding space for similarity estimation. However, the large semantic gap between the visual and textual modalities still hampers effective moment-query alignment. Additionally, fine-grained interactions, such as frame-word interactions, are often under-explored.
2. **Multiple Instance Learning (MIL)**: MIL has been widely adopted [9, 21] to enforce video-query alignment in the absence of moment-level supervision. However, many moments in a video are irrelevant to the query. Aggregating all moment scores can introduce noisy information, which negatively impacts the model's performance.

To overcome these limitations, we propose a Siamese Alignment Network (SAN) for weakly-supervised video moment retrieval. As illustrated in Fig. 6.2, our approach consists of two key components: a multi-scale Siamese module and a context-aware MIL module. The multi-scale Siamese module progressively reduces the semantic gap between the visual and textual modalities by projecting their representations into a shared embedding space, improving moment-query alignment. The context-aware MIL module enhances video-query alignment by considering contextual moments, thus reinforcing the weak supervision signal. Our experimental results on two benchmark datasets, Charades-STA [8] and ActivityNet Captions [13], demonstrate the effectiveness and superiority of our approach over

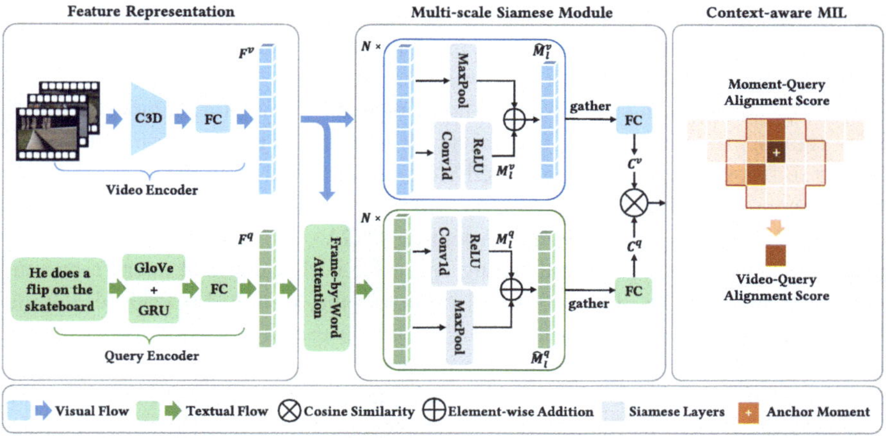

Fig. 6.2 The pipeline of our proposed Siamese Alignment Network (SAN). The multi-scale Siamese module generates moment candidates at multiple scales and aligns them with the query, while the context-aware Multiple Instance Learning (MIL) module refines the video-query matching by considering contextual moments for improved alignment

several state-of-the-art baselines. To further support research in this area, we also release our code.[1]

6.2 Related Work

This work is closely related to the areas of video moment localization, Siamese networks, and multiple instance learning.

Video Moment Localization Video moment localization aims to identify and retrieve specific moments within a video based on a given query. Existing methods can be broadly classified into two categories: supervised and weakly-supervised approaches.

Supervised methods are typically divided into two-stage, one-stage, and reinforcement learning approaches. In the two-stage setting, models generate moment candidates in an offline manner, such as by using multi-scale sliding windows [8, 17] or designing proposal generation networks [4, 37, 39]. One-stage methods, in contrast, directly retrieve the target moment in a single pass, as exemplified by Chen et al. [5], who proposed a boundary model to predict the start and end points of the moment. Although one- and two-stage methods have shown progress, they often suffer from inefficiency and lack flexibility. Reinforcement learning

[1] https://sancode.wixsite.com/san-model.

approaches, such as the work by He et al. [10], treat video moment retrieval as a sequential decision-making problem and introduce an end-to-end reinforcement learning framework for this task.

Despite advances in supervised methods, manually annotating temporal boundaries remains time-consuming and prone to inconsistency. As a result, there has been a shift towards weakly-supervised video moment retrieval, where only video-level annotations are available (e.g., the video associated with the query). Mithun et al. [21] introduced Text-Guided Attention (TGA), the first weakly-supervised model, which learns a joint embedding space for both video and query representations. Following this, Gao et al. [9] proposed the Weakly Supervised Natural Language Localization Network (WSLLN), which uses a two-stream structure to measure moment-query consistency. Despite these promising results, many weakly-supervised methods still rely on two-stage approaches with multi-scale sliding windows, leading to inefficiencies. To address this, Lin et al. [15] introduced the Semantic Completion Network (SCN), which scores all moments in a single pass. Wu et al. [36] proposed the Boundary Adaptive Refinement (BAR) framework, which refines temporal boundaries using reinforcement learning. More recently, Wang et al. [35] developed the Weakly Supervised Temporal Adjacent Network (WSTAN), which models temporal relations using a 2D feature map. While these methods have improved weakly-supervised video moment retrieval, significant challenges remain due to the large semantic gap between the visual and textual modalities.

Siamese Network The Siamese network, initially introduced by Bromley et al. [1] for signature verification, consists of two or more identical sub-networks that share parameters and are used to measure the similarity between instances. In the image domain, Chopra et al. [7] applied a Siamese network for face verification, while Varior et al. [32] used a deep Siamese CNN architecture for human re-identification, aiming to project similar pairs closer together and dissimilar pairs farther apart. In video analysis, Ryoo et al. [27] employed a multi-Siamese CNN to identify activities in low-resolution videos, and Tao et al. [30] developed a Siamese instance search tracker for target matching across frames. In the language domain, Mueller et al. [22] proposed a Siamese adaptation of the LSTM network for assessing semantic similarity between variable-length sentences.

While Siamese networks have found widespread application in other domains, their use in weakly-supervised video moment retrieval has been limited. In this work, we leverage a multi-scale Siamese module to enhance visual and textual alignment, addressing both the challenges of semantic discrepancy and weak supervision in video moment retrieval.

Multiple Instance Learning MIL was introduced by Keeler et al. [12] for handwritten digit recognition. In MIL, a set of elements is treated as a "bag" and is associated with weak supervision, where only set-level labels are available instead of individual instance-level labels. This approach enables models to predict group-level information despite the absence of detailed supervision for individual instances, making MIL an attractive option for real-world problems with challenging labeling.

In standard MIL, a set is considered positive if it contains at least one positive instance, and negative if all instances in the set are negative. Wang et al. [33] used a neural network to learn a set representation and perform set classification without estimating individual instance-level probabilities. Ilse et al. [11] introduced an attention mechanism to perform pooling over instances, where the attention weights learned from the instances reflect each instance's contribution to the final decision. Similarly, Liu et al. [16] used computational set representation to measure distances between sets of images, while Yan et al. [38] iteratively updated the contributions of instances over multiple iterations.

Although MIL has been applied to weakly-supervised video moment retrieval, many methods do not fully optimize MIL for the specific challenges of this task. A common issue is the aggregation of all moment scores to compute a video-query matching score. This can lead to the inclusion of noisy, query-unrelated moments, adversely affecting retrieval performance. To address this, we propose incorporating Intersection over Union (IoU) scores as prior knowledge to mitigate the influence of query-unrelated moments, thereby improving the overall retrieval accuracy.

6.3 Methodology

As illustrated in Fig. 6.2, our proposed SAN comprises the following components: (1) Feature Representation Component: This component encodes both visual and textual information, projecting the resulting embeddings into a joint embedding space for unified representation. (2) Multi-Scale Siamese Module: This module generates multi-scale moment candidates online and enhances fine-grained cross-modal semantic alignment between visual and textual modalities. (3) Context-Aware MIL Module: This module leverages adjacent moment candidates to facilitate the estimation of video-query similarity, ensuring more accurate moment localization.

6.3.1 Problem Formulation

Let $\mathcal{V}$ and Q denote a video and a query, respectively. We represent a video as a sequence of frames $\mathcal{V} = \{v_i\}_{i=1}^{n_v}$, where v_i represents the i-th video frame, and n_v denotes the total number of frames. Similarly, we represent the query as a sequence of words $Q = \{q_i\}_{i=1}^{n_q}$, where q_i denotes the i-th word and n_q is the number of words in the query. The query is associated with a temporal annotation $\tau = \{t_s, t_e\}$, where t_s and t_e represent the start and end points of the target video moment, respectively.

Our goal is to identify a golden moment $c = \{v_t\}_{t=\hat{t}_s}^{\hat{t}_e}$ in the video $\mathcal{V}$ that corresponds to the given query Q under the weakly-supervised setting, such that

$(\hat{t}_s, \hat{t}_e) = (t_s, t_e)$. In this weakly-supervised scenario, temporal annotations are unavailable during training, though they are accessible for evaluation.

Formally, the weakly-supervised video moment retrieval task can be defined as follows:

Input: A training set consisting of untrimmed videos and their corresponding queries, without temporal annotations.

Output: A retrieval model that generates moment candidates and estimates the relevance score of each moment-query pair.

6.3.2 Feature Representation

Given the untrimmed video $\mathcal{V} = \{v_i\}_{i=1}^{n_v}$, we adopt the pre-trained C3D network [31] to extract frame-level features, as done in [21]. Specifically, we extract frame features from the penultimate fully connected (FC) layer of the C3D network. These features are then sampled to a fixed length using linear interpolation. Subsequently, we apply an FC layer to project the frame features into a joint embedding space, resulting in the final frame representations, denoted as $\mathbf{F}^v = \{\mathbf{f}_i^v\}_{i=1}^{n_v}$, where $\mathbf{f}_i^v \in \mathbb{R}^d$ represents the final representation of the i-th frame and d is the dimension of the representation.

For the query $\mathcal{Q} = \{q_i\}_{i=1}^{n_q}$, we first use the pre-trained GloVe [25] embeddings to extract the initial word features. These word embeddings are then passed through a Gated Recurrent Unit (GRU) network [6] to capture contextual information for each word in the query. Finally, we project the context-aware word features into the joint embedding space using an FC layer with Batch Normalization and Dropout, resulting in the final query representations $\mathbf{F}^q = \{\mathbf{f}_i^q\}_{i=1}^{n_q}$, where $\mathbf{f}_i^q \in \mathbb{R}^d$.

6.3.3 Multi-Scale Siamese Module

Most existing methods, such as [21] and [9], adopt the multi-scale sliding window strategy to generate moment candidates offline, lacking flexibility. Although these approaches improve semantic alignment between the moment and the description query, the heterogeneity gap remains a challenging issue. To address this, we introduce the Multi-scale Siamese Module (MSM), which consists of two branches: a visual branch and a textual branch, as illustrated in Fig. 6.2. Below, we describe the functionality of each branch.

6.3.3.1 Visual Branch

Our task is to identify the golden moment corresponding to a given query description. To achieve this, we first need to generate moment candidates. We introduce a hierarchical convolution module, consisting of multiple one-dimensional convolution layers. Each convolution layer outputs a set of video moment candidates along with their corresponding representations. To improve consistency during training and enhance the representation of each candidate, we incorporate residual mappings on top of each convolution layer. The process can be formulated as follows:

$$\begin{cases} \mathbf{M}_l^v = ReLU(Conv1d(\hat{\mathbf{M}}_{l-1}^v, \theta_l^k, \theta_l^s)), \\ \hat{\mathbf{M}}_l^v = \mathbf{M}_l^v + MaxPool(\hat{\mathbf{M}}_{l-1}^v, \omega_l^k, \omega_l^s), \end{cases} \tag{6.1}$$

where θ_l^k and θ_l^s are the kernel size and stride size of the l-th convolution layer, respectively, and ω_l^k and ω_l^s are the kernel size and stride size of the l-th MaxPool layer. $\hat{\mathbf{M}}_l^v$ is the final output of the l-th layer, and $\hat{\mathbf{M}}_{l-1}^v$ is the input feature matrix at layer l ($l \geq 1$), with $\hat{\mathbf{M}}_0^v = \mathbf{F}^v$ being the input to the first layer.

Each convolution layer has a different receptive field, and $\hat{\mathbf{M}}_l^v$ stores moment candidates with different temporal boundaries. The temporal boundaries of these moment candidates can be adjusted by varying θ_l^k and θ_l^s. The temporal boundaries for each moment candidate are estimated as follows:

$$\begin{cases} w_l = \prod_{i=1}^{l} \theta_i^s, \\ r_l = r_{l-1} + (\theta_l^k - 1) * w_{l-1}, \\ t_{l,j}^s = w_l * j, \\ t_{l,j}^e = t_{l,j}^s + r_l, \end{cases} \tag{6.2}$$

where w_l is the accumulated stride size of the previous l layers, r_l is the receptive field size of the l-th layer, and $t_{l,j}^s$ and $t_{l,j}^e$ represent the start and end time points of the j-th moment in the l-th layer.

After the hierarchical convolution module, we obtain the representations of moment candidates and their corresponding temporal boundaries. These moment candidates are gathered from all layers, and passed through an FC layer with a ReLU activation function to produce the final moment candidate representations, denoted as $C^v = \{\mathbf{m}_i^v\}_{i=1}^{n_m}$, where $\mathbf{m}_i^v$ represents the i-th moment candidate and n_m is the number of moment candidates. The corresponding temporal boundaries are denoted as $\mathcal{T} = \{(t_i^s, t_i^e)\}_{i=1}^{n_m}$.

6.3.3.2 Textual Branch

To enhance the fine-grained semantic alignment between the moment candidates and the query, we introduce a Siamese structure. Specifically, we employ an identical hierarchical convolution module to process the query, sharing parameters with the visual branch. This setup allows the model to capture aligned semantics between the two branches, further reducing the semantic gap.

We first apply a frame-by-word attention mechanism to transfer the textual information into the visual space, generating frame-specific textual representations. This process ensures that the input to the textual branch has the same shape as the visual branch. The attention mechanism can be described as follows:

$$\begin{cases} \hat{\mathbf{f}}_i^q = \sum_{j=1}^{n_q} a_{i,j}\mathbf{f}_j^q, \\ a_{i,j} = \frac{exp(z_{i,j})}{\sum_{k=1}^{n_q} exp(z_{i,k})}, \\ z_{i,j} = \mathbf{f}_i^{vT}\mathbf{f}_j^q, \end{cases} \tag{6.3}$$

where $a_{i,j}$ is the normalized attention weight of the i-th frame with respect to the j-th word, and $\hat{\mathbf{f}}_i^q$ is the textual representation corresponding to the i-th visual frame. As a result, the sequence of textual representations has the same length as the sequence of visual frames, with each textual representation semantically associated with its corresponding frame representation.

Next, we pass the sequence of textual representations through an identical hierarchical convolution network, obtaining moment-specific textual representations as follows:

$$\begin{cases} \mathbf{M}_l^q = ReLU(Conv1d(\hat{\mathbf{M}}_{l-1}^q, \theta_l^k, \theta_l^s)) \\ \hat{\mathbf{M}}_l^q = \mathbf{M}_l^q + MaxPool(\hat{\mathbf{M}}_{l-1}^q, \omega_l^k, \omega_l^s), \end{cases} \tag{6.4}$$

where $\hat{\mathbf{M}}_{l-1}^q$ is the input to the l-th layer, and $\hat{\mathbf{M}}_0^q = \hat{\mathbf{F}}^q$ is the input to the first layer.

After processing the textual branch, we gather the moment-specific textual representations from all layers, feed them into an FC layer with a ReLU activation function (sharing parameters with the visual branch), and obtain the final textual representations of the moment candidates, denoted as $C^q = \{\mathbf{m}_i^q\}_{i=1}^{n_m}$.

By leveraging the Siamese structure, our model effectively enhances the semantic alignment between the visual and textual modalities, ensuring consistent semantics across different granularities. With the moment representations C^v and C^q, we compute the moment-query alignment score using cosine similarity as follows:

$$s_i = \frac{\mathbf{m}_i^{vT}\mathbf{m}_i^q}{\|\mathbf{m}_i^v\|\|\mathbf{m}_i^q\|}. \tag{6.5}$$

6.3.4 Context-Aware MIL

To improve retrieval performance by considering the video-level annotations, we propose a Context-aware MIL (C-MIL) module. This module enhances the alignment between moment-query pairs and video-query pairs by incorporating contextual moment candidates.

6.3.4.1 Enhanced Moment-Query Alignment

Previous works estimate the alignment score of each moment-query pair by simply measuring the similarity between their representations. In contrast, we aggregate the alignment scores of contextual moments to strengthen the alignment score of the current moment. This approach is based on the idea that moments with higher temporal IoU values provide complementary information. As illustrated in Fig. 6.3, compared to Candidate 2, Candidate 1 has a higher IoU value with the anchor moment. Moreover, Candidate 1's visual information is more similar to that of the anchor, contributing to a better understanding of the anchor. In other words, the alignment scores of contextual moments can supplement the alignment score of the current moment.

Inspired by this, we first calculate the IoU score between a pair of moment candidates as follows:

$$\begin{cases} I_{i,j} = max[min(t_i^e, t_j^e) - max(t_i^s, t_j^s), 0], \\ U_{i,j} = (t_i^e - t_i^s) + (t_j^e - t_j^s) - I_{i,j}, \\ \delta_{i,j} = \frac{I_{i,j}}{U_{i,j}}, \end{cases} \tag{6.6}$$

where $I_{i,j}$ and $U_{i,j}$ represent the intersection and union between the i-th and j-th moment candidates, respectively. Here, t_i^s and t_i^e denote the start and end points of the i-th moment candidate, and t_j^s and t_j^e represent the start and end points of the j-th moment candidate.

Subsequently, we strengthen the alignment estimation of each moment-query pair by considering its contextual moments as follows:

$$\begin{cases} \alpha_{i,j} = \frac{exp(\delta_{i,j})}{\sum_{k=1}^{n_c} exp(\delta_{i,k})}, \\ \hat{s}_i = s_i + \lambda \sum_{j=1}^{n_c} \alpha_{i,j} s_j, \end{cases} \tag{6.7}$$

where s_i is the original alignment score of the i-th moment, $\hat{s}_i$ is the enhanced score, λ is a hyper-parameter that controls the balance between the original score and the

Fig. 6.3 Illustration of the contribution of contextual moments for enhanced moment-query alignment

contextual contributions, and n_c is the number of moment candidates excluding the i-th moment (i.e., $n_c = n_m - 1$).

6.3.4.2 Video-Query Alignment

To estimate the matching score between each video and the corresponding query, we treat the video as a "bag" and each moment candidate within the video as an instance of this "bag". A straightforward approach to obtain the bag-level alignment score is to select the highest score among all instances in the bag. However, under a weakly-supervised setting, this approach can be problematic—especially early in training—because the model's learning ability is initially limited, leading to potentially inaccurate predictions. Simply selecting the highest-scoring instance may thus introduce negative influences to the model. Although previous work [9, 21] attempts to alleviate this issue by aggregating scores across all instances, treating all instances equally often introduces noise, negatively impacting the learning process.

To overcome these challenges, we propose a novel video-query alignment score estimation strategy. Specifically, we first select the highest-scoring moment candidate as an anchor, identifying the most relevant instance from the "bag". We then leverage IoU scores, relative to the anchor, as guidance information to assess the confidence of each moment candidate.

More concretely, we first calculate the IoU scores between the anchor moment and the other moment candidates as follows:

$$\begin{cases} I_i = max[min(t^e, t_i^e) - max(t^s, t_i^s), 0], \\ U_i = (t^e - t^s) + (t_i^e - t_i^s) - I_i, \\ \delta_i = \frac{I_i}{U_i}, \end{cases} \tag{6.8}$$

where I_i and U_i represent the intersection and union between the i-th moment candidate and the anchor moment, respectively. Here, t^s and t^e are the start and end points of the anchor moment, while t_i^s and t_i^e are the start and end points of the i-th moment candidate.

Subsequently, we estimate the alignment score of the video-query pair by performing a weighted aggregation of the candidate scores, guided by their IoU values:

$$\begin{cases} \alpha_i = \frac{exp(\delta_i)}{\sum_{k=1}^{n_c} exp(\delta_k)}, \\ S = \hat{s} + \mu \sum_{i=1}^{n_c} \alpha_i \hat{s}_i, \end{cases} \tag{6.9}$$

where $n_c = n_m - 1$ represents the number of moment candidates excluding the anchor, $\hat{s}$ is the strengthened alignment score of the anchor moment, S is the final alignment score for the given video-query pair, and μ is a balance hyper-parameter.

6.3.5 Learning

To achieve semantic alignment between a given positive video-query pair (V, Q), we optimize the model using a margin-based ranking loss, which is defined as:

$$L = \sum_{(V,Q)} \{\sum_{V^-} max[0, \Delta - S(V, Q) + S(V^-, Q)] + \sum_{Q^-} max[0, \Delta - S(Q, V) + S(Q^-, V)]\}, \quad (6.10)$$

where Δ is the margin, and (V^-, Q) and (Q^-, V) represent negative pairs. Specifically, (Q^-, V) is constructed by selecting a query Q^- from a different video in the mini-batch, while (V^-, Q) is formed by selecting a video V^- from a different query in the mini-batch.

6.4 Experiments

In this section, we conducted experiments on two benchmark datasets to evaluate the effectiveness of our proposed model. We began by introducing the datasets and the evaluation protocols, followed by a description of the implementation details. We then presented the results on the Charades-STA [8] and ActivityNet-Captions [13] datasets. Subsequently, we provided an analysis of the contributions of the Siamese network and the C-MIL module. Finally, we presented several qualitative results to further demonstrate the model's performance.

6.4.1 Experimental Settings

6.4.1.1 Datasets

We evaluated the performance of the proposed SAN model using two benchmark datasets: Charades-STA and ActivityNet-Captions.

Charades-STA [8] It was initially proposed by Sigurdsson et al. [28], containing action interval annotations and video-level language descriptions. Later, Gao et al. [8] extended the annotations for video moment retrieval by decomposing video-level descriptions into short sentences and assigning these sentences to specific video clips based on activity category information. The average query length is 8.6 words, and the average video duration is 29.8 seconds. The Charades-STA dataset includes a total of 16,128 video-query pairs, with 12,408 used for training and the remaining 3720 for testing.

ActivityNet-Captions [13] It is currently the largest dataset for video moment retrieval, containing 19,209 videos and 17,031 language queries, each with corresponding temporal annotations. In this paper, we follow the standard setup used in previous works, using val_1 as the validation set and val_2 as the testing set. The average query length is 13.16 words, and the average video duration is 117.74 seconds.

6.4.1.2 Evaluation Protocols

To comprehensively evaluate our model and compare it with baselines, we used the "R@n, IoU=m" metric, as employed in [21]. Specifically, for a given query, this metric measures the percentage of the top-n retrieved results that have an IoU greater than m. Formally, it is defined as:

$$R@n, IoU = m = \frac{1}{k}\sum_{i=1}^{k} r(n, m, q_i), \tag{6.11}$$

where $r(n, m, q_i)$ represents the recall rate for query q_i, and k is the total number of queries.

6.4.1.3 Implementation Details

We optimized our proposed model using PyTorch on a TITAN XP GPU. The Adam optimizer was used for 15 epochs, with a learning rate initialized at 5e-5, adjusted using cosine annealing [20] to a final learning rate of 5e-7. The batch size was set to 32 for both Charades-STA and ActivityNet-Captions datasets. To ensure uniform sequence lengths across videos in each dataset, we used linear interpolation. Specifically, the sequence length was set to 31 for Charades-STA and 127 for ActivityNet-Captions. For the Charades-STA dataset, the multi-scale Siamese module consists of 3 layers, with all kernel sizes set to 3 and stride sizes to 2. For ActivityNet-Captions, the corresponding values are 4 layers, kernel size of 7, and stride size of 2. For both datasets, the dimension of the joint embedding space is set to 512. Additionally, all FC layers and convolution layers in the multi-scale Siamese module have an output dimension of 512. The margin for the triplet loss is set to 0.2 for Charades-STA and 0.1 for ActivityNet-Captions. The hyper-parameters λ and μ are set to 0.7 and 0.3 for Charades-STA, and 0.7 and 1.0 for ActivityNet-Captions, respectively. Note that moment-query alignment scores are strengthened only during the training phase.

During inference, we first passed the video-query pairs through the SAN model to obtain moment-query alignment scores. Next, we applied soft Non-Maximum Suppression (NMS) to filter out redundant moment candidates. Finally, the remaining candidates are ranked to identify the target moment.

6.4.1.4 Baselines

Following [35], we compared our proposed SAN model with the following state-of-the-art baselines to validate its effectiveness:

- **TGA** [21]: This method first generates moment candidates using sliding windows, then introduces a text-guided attention module to obtain attention weights, which are used to calculate query-specific video features. Both video and query representations are projected into a joint embedding space through two FC layers.
- **WSLLN** [9]: This method is a two-branch network comprising an alignment branch and a detection branch, trained using pseudo segment-level labels.
- **SCN** [15]: This method scores all sampled moments at different scales in a single pass. It employs a semantic completion module to measure semantic similarity between moments and the query. Additionally, it uses a reconstruction loss to train the semantic completion module, encouraging the model to focus on key visual information.
- **BAR** [36]: This approach utilizes reinforcement learning to progressively refine the temporal boundaries of moments. It also employs a MIL-based alignment evaluator to measure the alignment score of each moment-query pair, providing tailor-designed rewards.
- **WSTAN** [35]: It learns cross-modal semantic alignment by exploiting a temporal adjacent network, using a complete description paragraph as input.
- **Supervised Methods**: For a more comprehensive comparison, we also included several representative supervised learning methods, such as CTRL [8], ROLE [17], QSPN [37], 2D-TAN [39], DPIN [34], PFGA [23], FIAN [26], and CSMGAN [19].

Note that we directly quoted the results of these baselines from their original papers, as the source codes and associated parameters were not publicly released by the authors (except for TGA).

6.4.2 Performance Comparison

The comparison results on the Charades-STA and ActivityNet-Captions datasets are summarized in Tables 6.1 and 6.2, respectively. From analyzing these tables, we made the following observations:

- TGA demonstrates poor performance on Charades-STA as it lacks the ability to model fine-grained interactions between visual and textual modalities. Additionally, WSLLN performed worst on ActivityNet-Captions, mainly because it aggregates scores from all moment-query pairs to determine the video-query alignment score, which introduces noise.
- SCN consistently outperforms TGA and WSLLN, which validates the importance of incorporating contextual information. BAR also outperforms TGA and

Table 6.1 Performance comparison between our proposed model and state-of-the-art baselines on the Charades-STA dataset. The best results are highlighted in bold

		R@1			R@5		
Category	Method	IoU = 0.7	IoU = 0.5	IoU = 0.3	IoU = 0.7	IoU = 0.5	IoU = 0.3
Supervised	CTRL [8]	8.89	23.63	–	29.52	58.92	–
	ROLE [17]	7.82	21.74	37.69	30.06	70.37	92.79
	QSPN [37]	15.80	35.60	54.70	45.40	79.40	**95.60**
	2D-TAN [39]	23.25	39.81	–	52.15	79.33	–
	DPIN [34]	26.96	47.98	–	55.00	85.53	–
	PFGA [23]	33.74	52.02	**67.53**	–	–	–
	FIAN [26]	**37.72**	**58.55**	–	**63.52**	**87.80**	–
Weakly-supervised	TGA [21]	5.73	15.13	27.74	25.22	56.83	83.23
	SCN [15]	9.97	23.58	42.96	38.87	71.80	**95.56**
	BAR [36]	12.23	27.04	44.97	–	–	–
	WSTAN [29]	12.28	29.35	43.39	41.53	**76.13**	93.04
	SAN	**13.12**	**31.02**	**51.02**	**41.75**	72.56	89.95

Table 6.2 Performance comparison between our proposed model and state-of-the-art baselines on the ActivityNet-Captions dataset. The best results are highlighted in bold

		R@1			R@5		
Category	Method	IoU = 0.7	IoU = 0.5	IoU = 0.3	IoU = 0.7	IoU = 0.5	IoU = 0.3
Supervised	QSPN [37]	13.60	27.70	45.30	38.30	59.20	75.70
	PFGA [23]	19.26	33.04	51.28	–	–	–
	2D-TAN [39]	27.38	44.05	58.75	**62.26**	76.65	85.65
	DPIN [34]	28.31	47.27	62.40	60.03	77.45	87.52
	CSMGAN [19]	29.15	**49.11**	**68.52**	59.63	77.43	**87.68**
	FIAN [26]	**29.81**	47.90	64.10	59.66	**77.64**	87.59
Weakly-supervised	WSLLN [9]	–	22.70	42.80	–	–	–
	SCN [15]	–	29.22	47.23	–	55.69	71.45
	BAR [36]	–	**30.73**	49.03	–	–	–
	WSTAN [29]	–	30.01	**52.45**	–	63.42	79.38
	SAN	**13.85**	30.54	48.44	**29.96**	**64.52**	**82.41**

WSLLN on both datasets, which can be attributed to its use of a tailored reinforcement learning paradigm to adaptively refine temporal boundaries and reduce the cross-modal semantic gap. WSTAN, which models temporal relations using a 2D feature map (with one dimension indicating the start time and the other indicating the end time), achieves the best performance among all baseline models on both datasets.

- Our proposed SAN achieves superior performance, showing competitive results compared to all baselines on both datasets. Specifically, compared to TGA, SCN, BAR, and WSTAN, SAN achieves relative gains in "R@1 IoU=0.7" of 48.42, 31.60, 7.28, and 6.84% on Charades-STA, respectively. On ActivityNet-Captions, SAN also improves by 15.86% compared to SCN in terms of "R@5, IoU=0.5".
- Compared to weakly supervised approaches, supervised methods yield significantly higher retrieval performance on both datasets due to the availability of more extensive supervision information. Nevertheless, SAN outperforms some of the supervised approaches. For example, on Charades-STA, SAN demonstrates an improvement of 47.58 and 67.78% compared to CTRL and ROLE, respectively, in terms of "R@1, IoU=0.7".

6.4.3 *Ablation Study*

In this section, we conducted several experiments on Charades-STA to further evaluate the effectiveness of our model. Specifically, we examined the impact of the multi-scale Siamese module on moment retrieval results and analyzed how the C-MIL module influences retrieval performance.

6.4.3.1 On Multi-Scale Siamese Module

To validate the effectiveness of the MSM described in Sect. 6.3.3, we experimented with two model variants:

- **SAN w/o Param**: A variant without shared parameters between the visual and textual branches.
- **SAN w/o MSM**: A variant with the textual branch removed from the multi-scale Siamese module. In this case, after generating the multi-scale moment candidates from the visual branch, we used the frame-by-word attention mechanism introduced in Section 3.3 to learn a visual-specific query representation for each candidate. We then directly applied cosine similarity to estimate the moment-query alignment score.

We evaluated these model variants on the Charades-STA dataset, and the component-wise comparison results are shown in Fig. 6.4. Based on these results, we observed the following:

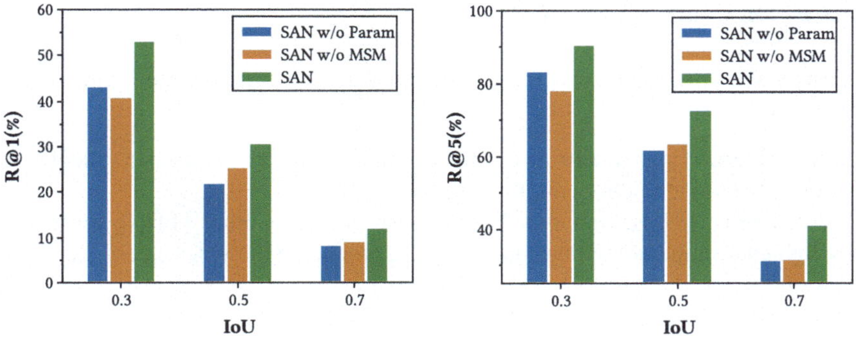

Fig. 6.4 Performance comparison of our model variants related to the MSM on the Charades-STA dataset

Table 6.3 Effects of the residual mapping on Charades-STA dataset

	R@1		
Method	IoU = 0.7	IoU = 0.5	IoU = 0.3
w residual	13.12	31.02	51.02
w/o residual	6.80	19.70	41.91

- Compared to our full model, the performance of SAN w/o Param shows a significant decline. Specifically, it drops by 18.66, 28.50, and 31.23% on R@1 metrics. This highlights the importance of parameter sharing between the visual and textual branches, which forms the basis of the Siamese structure. Sharing parameters helps enforce alignment between visual and textual information, leading to improved similarity estimation between the moment and the given query.
- Our model outperforms SAN w/o MSM, demonstrating that the Siamese structure effectively captures textual information related to the corresponding video moment, thereby boosting model performance.
- Our proposed model consistently outperforms all variants across all evaluation metrics, confirming the effectiveness and necessity of the multi-scale Siamese module for weakly supervised video moment retrieval.

As mentioned earlier, we added max pooling layers as residual mappings on top of each convolution layer to ensure consistent training. To evaluate the effect of the residual mapping, we tested a variant of the MSM module without the residual connections. As shown in Table 6.3, removing the residual mapping resulted in a 6.32% drop in "R@1, IoU=0.7". This indicates that the residual mapping plays a crucial role in maintaining consistency during training and significantly contributes to the model's performance.

6.4.3.2 On Context-Aware MIL

To verify the effectiveness of the C-MIL module, we experimented with several variants of our model:

- **SAN-Mean**: This variant uses the average of all moment-query scores to represent the video-query alignment score.
- **SAN-LSE**: This variant uses LogSumExp (LSE) pooling [14] to perform a weighted average of moment-query scores, producing the video-query alignment score.
- **SAN-Max**: This variant directly selects the highest moment-query score as the video-query matching score.

We evaluated these model variants on the Charades-STA dataset, and the component-wise comparison results are shown in Fig. 6.5. From the results, we made the following observations:

- SAN outperforms SAN-Mean by a significant margin across all metrics. This result shows that simply averaging the moment-query scores is insufficient for representing the video-query alignment score. Average aggregation assumes that all moments are linearly independent and contribute equally to the final relevance estimation, which fails to account for the varying importance of individual moments and struggles to suppress irrelevant or noisy information. The improvements achieved by SAN validate the effectiveness of C-MIL.
- Although SAN-LSE uses a weighted aggregation approach to obtain the video-query similarity score, it produces results similar to SAN-Mean. This could be due to the use of the probability distribution of scores as weights, which sometimes assigns equal weight to visually dissimilar and query-unrelated moments. The performance drop of SAN-LSE highlights the importance of incorporating contextual information to enhance the video-query alignment.

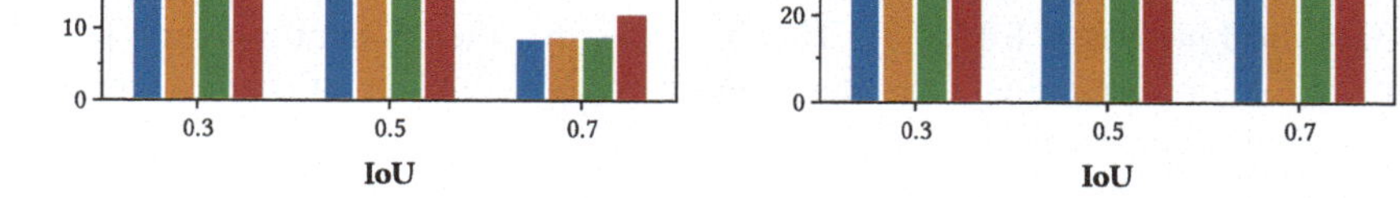

Fig. 6.5 Performance comparison among our model variants related to the C-MIL module on the Charades-STA dataset

Table 6.4 Ablation study of the C-MIL module in TGA on the Charades-STA dataset

Methods	R@1		
	IoU = 0.7	IoU = 0.5	IoU = 0.3
TGA	5.73	15.13	27.74
TGA-CMIL	6.21	16.10	29.54

- SAN-Max shows a performance decline, which reveals that relying solely on the highest-scored moment candidate is inadequate. In the weakly-supervised setting, the model's learning ability is initially limited, often resulting in imprecise predictions. Consequently, simply selecting the highest-scoring instance may have a negative impact on model training.
- Our SAN model consistently outperforms all variants on the Charades-STA dataset, demonstrating the critical importance of incorporating contextual information. Specifically, in our work, we use contextual information to enhance both moment-query and video-query alignment. These results strongly validate the effectiveness and necessity of our proposed C-MIL module.

To further verify the effect of the C-MIL module, we evaluated a variant of TGA on the Charades-STA dataset, named TGA-CMIL. Specifically, TGA uses a text-guided attention module to weight all candidate features and derive a global video feature. In TGA-CMIL, we replaced this approach by weighting only the highest-scored moment and adjacent moments with non-zero IoU scores relative to it. As shown in Table 6.4, TGA-CMIL achieves an 8.4% relative improvement in "R@1, IoU=0.7" on Charades-STA, proving the effectiveness of including contextual information. The slightly lower performance of TGA can be attributed to its approach of aggregating information from all candidates to estimate video-query similarity, which introduces noise. In contrast, the C-MIL module leverages IoU scores as prior knowledge to mitigate the influence of query-unrelated moments.

6.4.4 Parameter Analysis

In this section, we explored the influence of four pivotal parameters on the performance of our model.

6.4.4.1 On Kernel Size

We first investigated the influence of kernel sizes in MSM on Charades-STA. Specifically, we initialized the kernel sizes of all layers to 3 based on empirical observations. We then tested four different kernel sizes for each layer by setting the kernel sizes of all preceding layers to the best verified values while keeping the remaining layers at their default values. In this analysis, we used "R@1, IoU=0.7" as the primary evaluation metric. The experimental results are shown in Fig. 6.6.

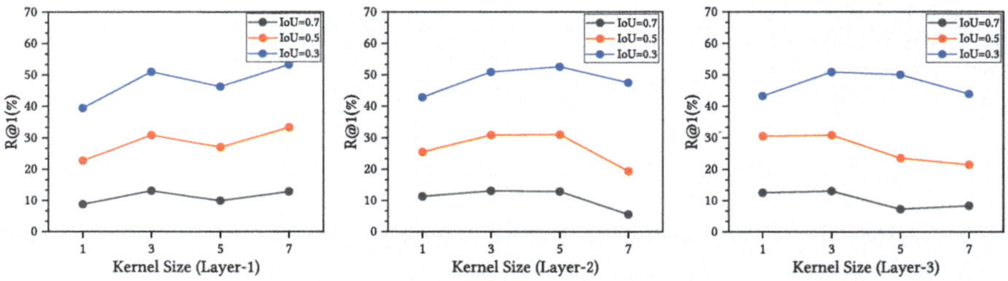

Fig. 6.6 Effect of kernel size in the multi-scale Siamese module. The first figure shows R@1 results for various kernel sizes of Layer-1, with Layer-2 and Layer-3 set to a kernel size of 3. The second figure presents R@1 results for Layer-2 with Layer-1 set to the optimal size from the first figure and Layer-3 set to 3. The final figure displays R@1 results for Layer-3, with Layer-1 and Layer-2 set to their optimal sizes from the previous figures

Table 6.5 Performance of our model with different layer numbers on Charades-STA dataset

	R@1		
Method	IoU = 0.7	IoU = 0.5	IoU = 0.3
3	13.12	31.02	51.02
2	12.07	30.97	43.79
1	1.34	8.52	31.59

From Fig. 6.6, we observed that setting all kernel sizes to 3 yields the best performance for "R@1, IoU=0.7". Moreover, SAN's performance fluctuates only slightly around this setting, indicating that SAN is not highly sensitive to kernel sizes near the optimal configuration.

6.4.4.2 On Layer Number

We also explored the effect of the number of layers in MSM. Since the stride size is set to 2, the maximum number of layers is limited to 3 on Charades-STA. The experimental results for different layer numbers are shown in Table 6.5. The results indicate that performance decreases sharply as the number of layers decreases. This is likely because reducing the number of layers reduces the number of candidates and the duration scales of moment candidates, thereby negatively affecting retrieval accuracy.

6.4.4.3 On Balance Parameters

We further explored the effect of the balance parameters λ and μ. Initially, we randomly set λ based on empirical experience and then explored different values for μ. After identifying the best value for μ, we then explored λ. Figure 6.7 shows the performance of our model with respect to these two parameters, achieved by varying one parameter while keeping the other fixed. We found that performance first

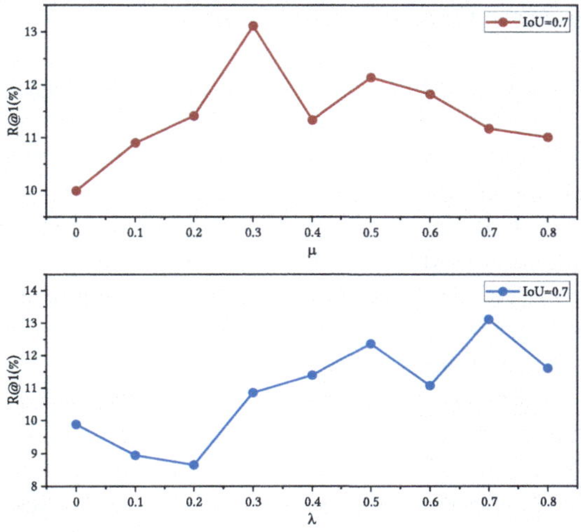

Fig. 6.7 Parameter tuning and sensitivity analysis of the balance hyper-parameters μ and λ on the Charades-STA dataset. The analysis was conducted by varying one parameter while keeping the other fixed

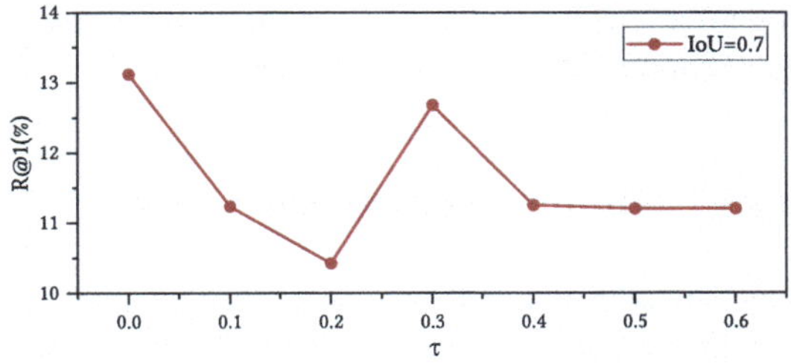

Fig. 6.8 Parameter tuning and sensitivity analysis of IoU threshold τ on Charades-STA

increases and then decreases as the parameters change, with the settings $\mu = 0.3$ and $\lambda = 0.7$ providing the best results on Charades-STA.

6.4.4.4 On IoU Threshold

In the C-MIL module, all candidates are considered as contextual moments to strengthen the video-query alignment. To further analyze the influence of the range of contextual moments, we introduced an IoU threshold τ. Specifically, we dropped candidates with IoU scores less than τ when calculating aggregation weights. As shown in Fig. 6.8, the performance exhibits a decreasing trend as τ increases, with the setting $\tau = 0$ (i.e., considering all candidates) yielding the best performance on Charades-STA.

6.4.5 *Visualization Results*

6.4.5.1 On Frame-by-Word Attention Weights

To validate the effectiveness of the frame-by-word attention mechanism, we visualized an example from the Charades-STA dataset in Fig. 6.9. As mentioned earlier, the sequence length of a video is 31 when fed into the frame-by-word attention module. For visualization purposes, we used a max pooling operation to downsample the attention weights, resulting in six weights for each word. Each row in Fig. 6.9 represents the attention weights of a video moment, with the central frame of each moment displayed as an overlay.

From Fig. 6.9, we can observe that the moment with the highest IoU score with the ground truth exhibits higher attention weights for the words "walks", "over" and "refrigerator", which helps in retrieving the query-related moment. Furthermore, the fourth moment, where the person walks back to the camera, also shows high attention weights for the word "walks". This indicates that the frame-by-word attention mechanism is adept at capturing person movements, likely benefiting from pre-training on action detection tasks.

6.4.5.2 On Moment Alignment Scores

One key advantage of SAN over other methods is the ability of its C-MIL module to enhance moment-query alignment. To demonstrate this, we selected a video-query

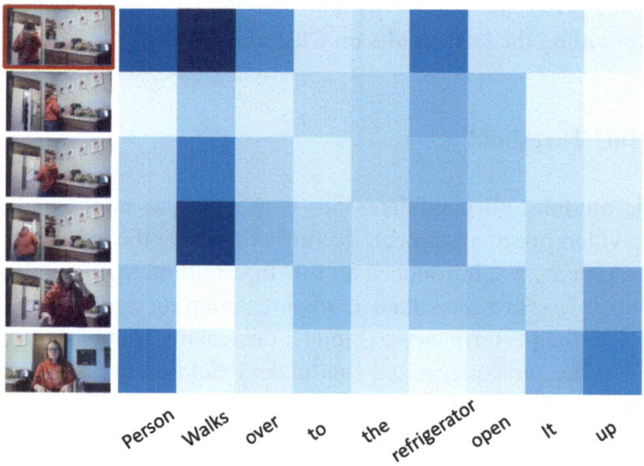

Fig. 6.9 Visualization of frame-by-word attention weights. Each row represents the weights of one down-sampled video moment across all words, while each column represents the weights of all moments for a specific word. Darker colors indicate higher weights. The moment with the highest IoU score with the ground truth is marked with a red box

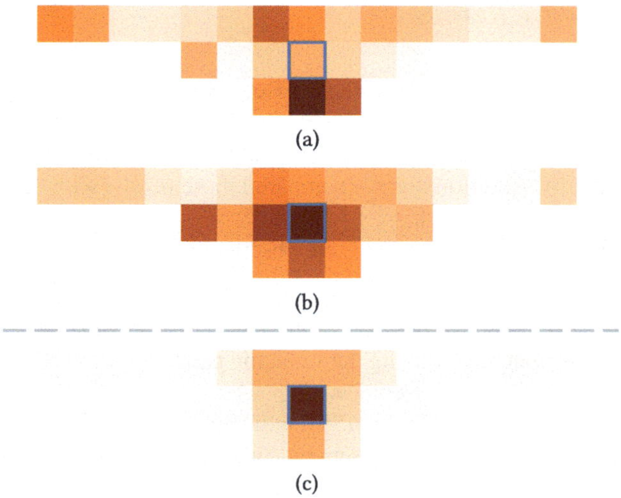

Fig. 6.10 (**a**) Initial moment alignment scores predicted by Eq. (6.5). (**b**) Context-enhanced moment alignment scores predicted by Eq. (6.7). (**c**) IoU-based weights of moment candidates with respect to the ground truth moment, marked in a blue box

pair from the Charades-STA dataset and visualized the moment-query alignment scores estimated by Eqs. (6.5) and (6.7) in Fig. 6.10.

From Fig. 6.10, we observe the following:

- Compared to the initial moment alignment scores in Fig. 6.10a, the context-enhanced moment alignment scores in Fig. 6.10b are more continuous in the temporal dimension, and the high-scored moments are more concentrated.
- For the anchor moment (i.e., the ground truth moment) shown in Fig. 6.10c, the enhancement primarily comes from its adjacent moments, which share similar and complementary visual information with the anchor. This helps boost the corresponding moment-query alignment score.
- The context-enhanced alignment scores for all moment candidates are diverse, suggesting that the aggregation of complementary information from contextual moments preserves the distinction between different moment candidates.

These findings align with our expectations and demonstrate that the C-MIL module effectively aggregates complementary contextual information without diminishing the individuality of moment candidates. This further validates the effectiveness of our C-MIL module.

6.4.6 Qualitative Results

To qualitatively validate the effectiveness of our SAN method, we presented two examples of video moment retrieval in Fig. 6.11. To analyze the effectiveness of each component, we also displayed the retrieval results of two variants: SAN-LSE and SAN w/o MSM. Additionally, we compared these results with the first weakly-supervised approach, TGA [21].

From the results shown in Fig. 6.11, we observed the following:

- Both SAN w/o MSM and SAN-LSE achieve better performance compared to TGA [21]. This demonstrates that both the MSM module and the context-aware MIL contribute to reducing the semantic gap between different modalities, thereby improving retrieval accuracy.
- SAN-LSE outperforms SAN w/o MSM, indicating that the MSM module plays a more significant role in enhancing semantic alignment between visual and textual modalities.
- Our complete SAN model achieves the best performance, significantly surpassing other baselines. This result highlights the importance of jointly incorporating both the context-aware MIL and the MSM module to enhance semantic alignment.

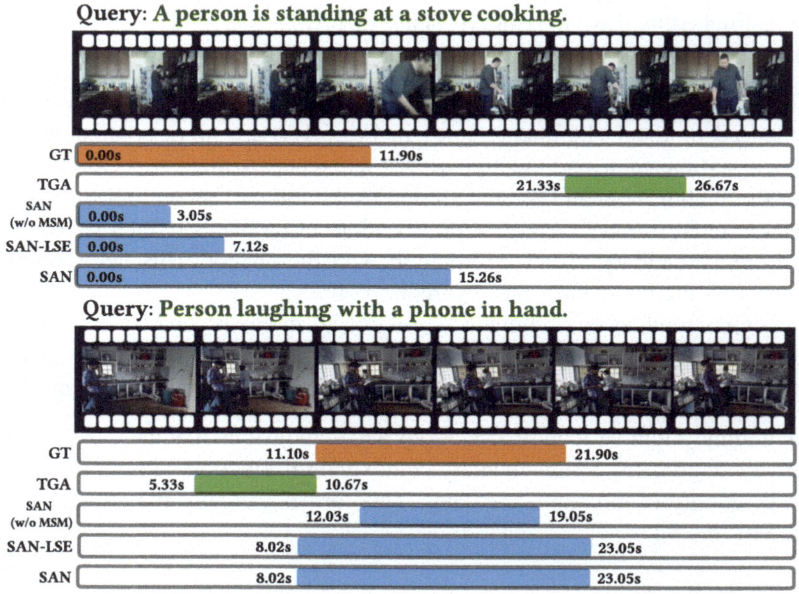

Fig. 6.11 Moment retrieval results on Charades-STA. All figures represent the R@1 results

6.5 Summary

In this chapter, we propose a Siamese alignment network for weakly-supervised video moment retrieval, which aims to identify the most relevant moment within a video based on a given natural language query. To effectively match moment candidates with the query, we design a multi-scale Siamese module that reduces the semantic gap between the visual and textual modalities, thus enhancing fine-grained semantic alignment. Additionally, we introduce a context-aware multiple-instance learning module to improve video-query alignment by considering the influence of adjacent moments. This dual approach enables more accurate video-query similarity estimation. Extensive experiments on two benchmark datasets validate the effectiveness of our method, demonstrating its superiority over several state-of-the-art baselines.

References

1. Bromley, J., Guyon, I., LeCun, Y., Säckinger, E., Shah, R.: Signature verification using a siamese time delay neural network. In: Proceedings of the Advances in Neural Information Processing Systems, pp. 737–744. Morgan Kaufmann, Burlington (1993)
2. Cao, D., Zeng, Y., Liu, M., He, X., Wang, M., Qin, Z.: STRONG: spatio-temporal reinforcement learning for cross-modal video moment localization. In: Proceedings of the International Conference on Multimedia, pp. 4162–4170. ACM, New York (2020)
3. Cao, D., Zeng, Y., Wei, X., Nie, L., Hong, R., Qin, Z.: Adversarial video moment retrieval by jointly modeling ranking and localization. In: Proceedings of the International Conference on Multimedia, pp. 898–906. ACM, New York (2020)
4. Chen, J., Chen, X., Ma, L., Jie, Z., Chua, T.: Temporally grounding natural sentence in video. In: Proceedings of the Conference on Empirical Methods in Natural Language Processing, pp. 162–171. ACL (2018)
5. Chen, J., Ma, L., Chen, X., Jie, Z., Luo, J.: Localizing natural language in videos. In: Proceedings of the American Association for Artificial Intelligence, pp. 8175–8182. AAAI (2019)
6. Cho, K., Merrienboer, B.V., Çaglar Gülçehre, Bahdanau, D., Bougares, F., Schwenk, H., Bengio, Y.: Learning phrase representations using RNN encoder–decoder for statistical machine translation. In: Proceedings of the Conference on Empirical Methods in Natural Language Processing, pp. 1724–1734. ACL (2014)
7. Chopra, S., Hadsell, R., LeCun, Y.: Learning a similarity metric discriminatively, with application to face verification. In: Proceedings of the Conference on Computer Vision and Pattern Recognition, pp. 539–546. IEEE, Piscataway (2005)
8. Gao, J., Sun, C., Yang, Z., Nevatia, R.: TALL: temporal activity localization via language query. In: Proceedings of the International Conference on Computer Vision, pp. 5277–5285. IEEE, Piscataway (2017)
9. Gao, M., Davis, L., Socher, R., Xiong, C.: WSLLN: weakly supervised natural language localization networks. In: Proceedings of the Conference on Empirical Methods in Natural Language Processing, pp. 1481–1487. ACL (2019)
10. He, D., Zhao, X., Huang, J., Li, F., Liu, X., Wen, S.: Read, watch, and move: reinforcement learning for temporally grounding natural language descriptions in videos. In: Proceedings of the American Association for Artificial Intelligence, pp. 8393–8400. AAAI (2019)

11. Ilse, M., Tomczak, J.M., Welling, M.: Attention-based deep multiple instance learning. In: Proceedings of the International Conference on Machine Learning, pp. 2132–2141. PMLR (2018)
12. Keeler, J.D., Rumelhart, D.E., Leow, W.K.: Integrated segmentation and recognition of hand-printed numerals. In: Proceedings of the Conference on Advances in Neural Information Processing Systems, pp. 557–563. Morgan Kaufmann, Piscataway (1990)
13. Krishna, R., Hata, K., Ren, F., Fei-Fei, L., Niebles, J.C.: Dense-captioning events in videos. In: Proceedings of the International Conference on Computer Vision, pp. 706–715. IEEE, Piscataway (2017)
14. Lee, K., Chen, X., Hua, G., Hu, H., He, X.: Stacked cross attention for image-text matching. In: Proceedings of the European Conference on Computer Vision, pp. 212–228. Springer, Berlin (2018)
15. Lin, Z., Zhao, Z., Zhang, Z., Wang, Q., Liu, H.: Weakly-supervised video moment retrieval via semantic completion network. In: Proceedings of the American Association for Artificial Intelligence, pp. 11539–11546. AAAI (2020)
16. Liu, Y., Yan, J., Ouyang, W.: Quality aware network for set to set recognition. In: Proceedings of the Conference on Computer Vision and Pattern Recognition, pp. 4694–4703. IEEE, Piscataway (2017)
17. Liu, M., Wang, X., Nie, L., Tian, Q., Chen, B., Chua, T.: Cross-modal moment localization in videos. In: Proceedings of the International Conference on Multimedia, pp. 843–851. ACM, New York (2018)
18. Liu, M., Wang, X., Nie, L., He, X., Chen, B., Chua, T.: Attentive moment retrieval in videos. In: Proceedings of the International SIGIR Conference on Research Development in Information Retrieval, pp. 15–24. ACM, New York (2018)
19. Liu, D., Qu, X., Liu, X., Dong, J., Zhou, P., Xu, Z.: Jointly cross- and self-modal graph attention network for query-based moment localization. In: Proceedings of the International Conference on Multimedia, pp. 4070–4078. ACM, New York (2020)
20. Loshchilov, I., Hutter, F.: SGDR: stochastic gradient descent with warm restarts. In: Proceedings of the International Conference on Learning Representations. OpenReview.net (2017)
21. Mithun, N.C., Paul, S., Roy-Chowdhury, A.K.: Weakly supervised video moment retrieval from text queries. In: Proceedings of the Conference on Computer Vision and Pattern Recognition, pp. 11592–11601. IEEE, Piscataway (2019)
22. Mueller, J., Thyagarajan, A.: Siamese recurrent architectures for learning sentence similarity. In: Proceedings of the American Association for Artificial Intelligence, pp. 2786–2792. AAAI (2016)
23. Opazo, C.R., Marrese-Taylor, E., Saleh, F.S., Li, H., Gould, S.: Proposal-free temporal moment localization of a natural-language query in video using guided attention. In: Proceedings of the Winter Conference on Applications of Computer Vision, pp. 2453–2462. IEEE, Piscataway (2020)
24. Peng, B., Lei, J., Fu, H., Jia, Y., Zhang, Z., Li, Y.: Deep video action clustering via spatio-temporal feature learning. Neurocomputing **456**, 519–527 (2021)
25. Pennington, J., Socher, R., Manning, C.D.: Glove: global vectors for word representation. In: Proceedings of the Conference on Empirical Methods in Natural Language Processing, pp. 1532–1543. ACL (2014)
26. Qu, X., Tang, P., Zou, Z., Cheng, Y., Dong, J., Zhou, P., Xu, Z.: Fine-grained iterative attention network for temporal language localization in videos. In: Proceedings of the International Conference on Multimedia, pp. 4280–4288. ACM, New York (2020)
27. Ryoo, M.S., Kim, K., Yang, H.J.: Extreme low resolution activity recognition with multi-siamese embedding learning. In: Proceedings of the American Association for Artificial Intelligence, pp. 7315–7322. AAAI (2018)
28. Sigurdsson, G.A., Varol, G., Wang, X., Farhadi, A., Laptev, I., Gupta, A.: Hollywood in homes: crowdsourcing data collection for activity understanding. In: Proceedings of the European Conference on Computer Vision, pp. 510–526. Springer, Berlin (2016)

29. Song, Y., Wang, J., Ma, L., Yu, Z., Yu, J.: Weakly-supervised multi-level attentional reconstruction network for grounding textual queries in videos. CoRR **abs/2003.07048** (2020). https://arxiv.org/abs/2003.07048
30. Tao, R., Gavves, E., Smeulders, A.W.M.: Siamese instance search for tracking. In: Proceedings of the Conference on Computer Vision and Pattern Recognition, pp. 1420–1429. IEEE, Piscataway (2016)
31. Tran, D., Bourdev, L.D., Fergus, R., Torresani, L., Paluri, M.: Learning spatiotemporal features with 3d convolutional networks. In: Proceedings of the International Conference on Computer Vision, pp. 4489–4497. IEEE, Piscataway (2015)
32. Varior, R.R., Haloi, M., Wang, G.: Gated siamese convolutional neural network architecture for human re-identification. In: Proceedings of the European Conference on Computer Vision, pp. 791–808. Springer, Berlin (2016)
33. Wang, X., Yan, Y., Tang, P., Bai, X., Liu, W.: Revisiting multiple instance neural networks. Pattern Recognit. **74**, 15–24 (2018)
34. Wang, H., Zha, Z., Chen, X., Xiong, Z., Luo, J.: Dual path interaction network for video moment localization. In: Proceedings of the International Conference on Multimedia, pp. 4116–4124. ACM, New York (2020)
35. Wang, Y., Deng, J., Zhou, W., Li, H.: Weakly supervised temporal adjacent network for language grounding. IEEE Trans. Multimedia **24**, 3276–3286 (2021)
36. Wu, J., Li, G., Han, X., Lin, L.: Reinforcement learning for weakly supervised temporal grounding of natural language in untrimmed videos. In: Proceedings of the International Conference on Multimedia, pp. 1283–1291. ACM, New York (2020)
37. Xu, H., He, K., Plummer, B.A., Sigal, L., Sclaroff, S., Saenko, K.: Multilevel language and vision integration for text-to-clip retrieval. In: Proceedings of the American Association for Artificial Intelligence, pp. 9062–9069. AAAI (2019)
38. Yan, Y., Wang, X., Guo, X., Fang, J., Liu, W., Huang, J.: Deep multi-instance learning with dynamic pooling. In: Proceedings of the Asian Conference on Machine Learning, pp. 662–677. PMLR (2018)
39. Zhang, S., Peng, H., Fu, J., Luo, J.: Learning 2d temporal adjacent networks for moment localization with natural language. In: Proceedings of the American Association for Artificial Intelligence, pp. 12870–12877. AAAI (2020)

Chapter 7
Efficient Hashing Based Video Moment Localization

7.1 Introduction

The exponential growth of video data has made retrieving specific moments from large video collections a prominent research topic in information retrieval [28, 33]. Traditional approaches primarily focus on retrieving entire videos that match a given query from large-scale datasets [4, 24]. However, videos often contain complex scenes with numerous objects and actions, where only specific segments are relevant to a given query, while the rest of the content may be irrelevant to end-users [20, 30]. For example, in the video shown in Fig. 7.1, a user may be interested in the specific moment when "the man fired at the white vehicle before fleeing," spanning from 30.5 to 72.8 s, even though the full video depicts a broader scene involving a robbery and escape. Therefore, retrieving precise moments from long, untrimmed videos using natural language queries, known as video moment localization, is increasingly valuable in real-world applications such as video surveillance [18, 19], robotic navigation [29], and autonomous driving [10].

Despite its growing importance, video moment localization faces several significant challenges, which can be broadly categorized as follows:

1. **Temporal Context Modeling**. In real-world queries, temporal references such as "first," "before," or "after" are common, making temporal context modeling crucial for accurate moment localization. Most existing methods [19, 35, 37] rely on recurrent neural networks like Bi-LSTM or Bi-GRU to encode contextual information. However, such methods struggle to effectively capture long-term dependencies and multi-scale semantics, especially in long video sequences [2], which limits their performance in accurately modeling temporal context.
2. **Moment Candidate Generation**. Many current approaches [1, 6, 18] employ sliding window strategies to divide videos into densely overlapping segments, treating these segments as moment candidates. While effective to some extent, this approach is computationally expensive and inefficient due to the NP-hard

M. Liu et al., *Intelligent and Efficient Video Moment Localization*,
https://doi.org/10.1007/978-3-031-87588-5_7

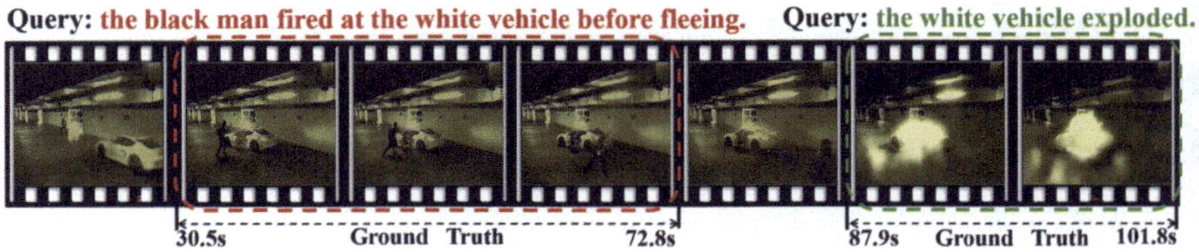

Fig. 7.1 An example of moment localization in a surveillance video

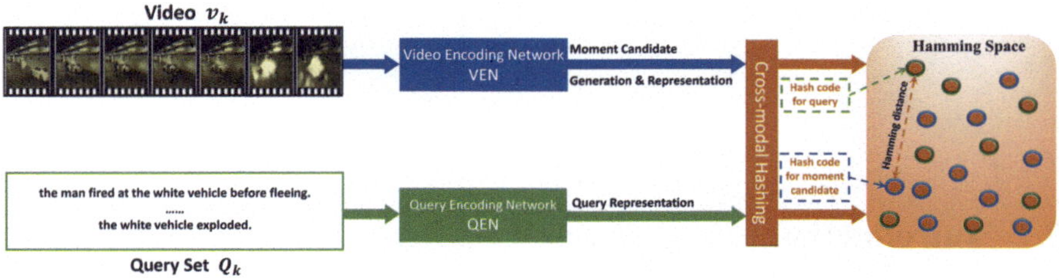

Fig. 7.2 Schematic illustration of our proposed CMHN approach, which seamlessly integrates the moment generation, representation, and localization within a unified model

nature of exhaustively searching through all possible segments. Additionally, offline moment generation networks cannot be trained end-to-end, limiting their scalability for large video datasets.

3. **Efficiency and Scalability**. In applications such as law enforcement, where multiple distinct moments from a surveillance video need to be localized with different queries (e.g., "the man fired at the white vehicle" and "the white vehicle exploded"), existing models [18, 19, 35, 37] rely on query-aware attention mechanisms. These "query-for-moment" models, however, repeat similar processes for each new query, even for the same video, leading to inefficiencies and scalability issues.

To address these challenges, we propose a novel end-to-end Cross-Modal Hashing Network (CMHN), illustrated in Fig. 7.2. Our method consists of a dual-path neural network with two independent components: the Video Encoding Network (VEN) and the Query Encoding Network (QEN). The VEN utilizes a Bidirectional Temporal Convolutional Network (Bi-TCN) to capture multi-scale temporal context information from the video, generating moment candidates along with enriched moment representations. Meanwhile, the QEN leverages Bidirectional Encoder Representations from Transformers (BERT) to capture the semantic content of the query and generate corresponding feature representations.

Once the video and query representations are acquired, a cross-modal hashing module projects these heterogeneous representations into a shared isomorphic Hamming space, facilitating compact hash code learning. The similarity between a video moment and a query is efficiently estimated using Hamming distance. One key advantage of our model is that once the hash codes for the video are learned, they can be precomputed offline, significantly improving the efficiency and scalability of the system in real-time localization scenarios. Additionally, we have made our code and parameter settings publicly available for research purposes.[1]

7.2 Related Work

This work is closely related to research in video moment localization and cross-modal hashing.

Video Moment Localization Video moment localization focuses on identifying specific segments within a video that are relevant to a given natural language query. Early efforts, such as those by Gao et al. [6, 7], address this problem by designing temporal unit regression networks that predict action proposals and refine temporal boundaries through coordinate regression. Around the same time, Hendricks et al. [1] proposed the Moment Context Network (MCN), which integrates local and global video features to facilitate query-based moment localization.

Recognizing that previous models struggled to capture the spatial-temporal relationships inherent in multi-modal data, Liu et al. [18, 19] developed two attention-based networks for moment localization. The first network enhances moment representations by capturing contextual information, while the second extracts important keywords from the query. Yuan et al. [35] introduced the Attention-Based Location Regression (ABLR) framework, which uses a co-attention memory model to capture spatial-temporal interactions between video segments and queries. This is followed by an attention-based regression network to generate the temporal coordinates of the target moment. To better model spatial-temporal information in both modalities, Xu et al. [32] proposed a multi-level model that fuses multi-modal data through early integration and utilizes video captioning as an auxiliary task to improve moment prediction accuracy.

Since prior research primarily focused on either contextual feature representation or spatial-temporal modeling, Zhang et al. [37] introduced the Cross-Modal Interaction Network (CMIN). CMIN employs graph convolution networks and multi-head self-attention to learn fine-grained representations, capturing intricate "frame-by-word" interactions across modalities. Furthermore, CMIN predicts alignment scores for moment candidates and adjusts the temporal boundaries of high-score moments to improve localization accuracy.

In summary, while significant progress has been made in learning joint representations through video-query interactions, existing methods often fall short in terms of retrieval efficiency and scalability. These models typically localize a single target moment for each query. To localize multiple moments in a video corresponding to different queries, they must process each query iteratively, which results in inefficiency and poor scalability. This "query-for-moment" processing

[1] The source code for CMHN can be accessed at https://github.com/Huyp777/CMHN.

strategy limits the practical applicability of these models, particularly for large-scale video datasets.

Cross-Modal Hashing Cross-modal hashing has gained popularity as an approach for retrieval tasks due to its advantages in storage efficiency and fast retrieval. Existing methods can be broadly classified into unsupervised and supervised approaches.

Unsupervised methods [3, 11, 25, 36] focus on learning hash functions by leveraging both intra- and inter-modality relationships. For example, Song et al. [25] introduced an inter-media hashing model that projects heterogeneous data into a shared Hamming space by co-regularizing intra- and inter-media consistency. Zhou et al. [36] proposed a latent semantic sparse hashing model that uses sparse coding and matrix factorization to capture high-level semantic information from images and text. Irie et al. [11] introduced an alternating co-quantization scheme that iteratively learns binary quantizers for each modality by jointly optimizing subspace learning and binary quantization. Moreover, Cui et al. [3] used social tags to enhance feature representation discrimination and improve hash code generation effectiveness.

Supervised methods [12, 16, 21, 26, 34] utilize semantic labels from training data to guide the learning of hash codes. For instance, Yu et al. [34] proposed a bi-stage discriminative coupled dictionary hashing model that jointly learns coupled dictionaries and hash functions for both modalities to capture underlying semantic information. Lin et al. [16] introduced a semantic-preserving hashing paradigm that minimizes the Kullback-Leibler divergence between probability distributions derived from semantic affinities to guide the hashing process. Many supervised methods, however, rely heavily on hand-crafted features, which results in a separation between feature extraction and hash code learning. To overcome this, Jiang et al. [12] developed an end-to-end cross-modal hashing framework based on deep neural networks, enabling direct learning of hash codes for each modality. Lu et al. [21] proposed a hierarchical recurrent network to exploit both spatial details and semantic information for effective hash code generation. In the context of fashion recommendation, Sun et al. [26] employed cross-modal hashing based on hierarchical label comprehension.

While substantial progress has been made in cross-modal retrieval for images, there has been limited exploration of hashing for video retrieval, which is often confined to mono-modal tasks such as near-duplicate video retrieval [8, 14]. Cross-modal hashing for video retrieval, especially for moment localization tasks, remains an underexplored research area.

7.3 Methodology

Our proposed CMHN, as illustrated in Fig. 7.2, consists of two main components: (1) a dual-path neural network architecture, comprising the Video Encoding Network (VEN) and the Query Encoding Network (QEN), and (2) a cross-modal hashing module designed to generate compact hash codes for both video moments and natural language queries. In this section, we detail the design and functionality of each component.

7.3.1 Problem Formulation

Consider a training video set consisting of N untrimmed videos, denoted as $\mathcal{S}_v = \{v_1, \ldots, v_k, \ldots, v_N\}$, where v_k represents the k-th video in the set. For each video v_k, there are M_k associated natural language queries, represented by $\mathcal{Q}_k = \{s_{k,1}, \ldots, s_{k,j}, \ldots, s_{k,M_k}\}$. Additionally, we define $\mathcal{A}_k = \{[t_s^{k,j}, t_e^{k,j}]\}_{j=1}^{M_k}$ as the set of ground truth temporal moments corresponding to the queries in $\mathcal{Q}_k$, where $[t_s^{k,j}, t_e^{k,j}] \in \mathcal{A}_k$ indicates the start and end times of the j-th target moment in v_k.

Given the training data, our objective is to learn a cross-modal hashing network. With the trained model, for a new untrimmed video v and its corresponding query set $\mathcal{Q}$, we first generate the moment candidate set C and compute the hash codes for each moment and query. Subsequently, we localize the target moments from C using Hamming distance, ensuring their timestamps align with the annotated ones.

7.3.2 The VEN Module

To generate the moment candidate set C containing moments of various lengths from the given video v and learn their representations, we propose a novel Video Encoding Network (VEN), as depicted in Fig. 7.3. VEN consists of two primary modules: temporal context modeling, and moment generation and representation.

7.3.2.1 Temporal Context Modeling

Given an untrimmed video v_k, we first employ a 3D convolutional network (C3D) [27] to extract the local feature sequence $\mathbf{X}_k = [\mathbf{x}_{k,1}, \ldots, \mathbf{x}_{k,i}, \ldots, \mathbf{x}_{k,R_x}]$, where $\mathbf{x}_{k,i}$ represents the i-th local feature of video v_k, and R_x denotes the length of the local feature sequence. To enhance these local features with contextual information, we build a model that considers both pre-context and post-context information. While existing methods predominantly use models like Bi-LSTM or Bi-GRU [15, 17] to capture contextual information from $\mathbf{X}_k$, these methods suffer from limitations. Specifically, as each hidden state only retains a portion of the information from its input and preceding hidden state, the resulting representation may not effectively capture long-term dependencies, especially for relatively long videos [5].

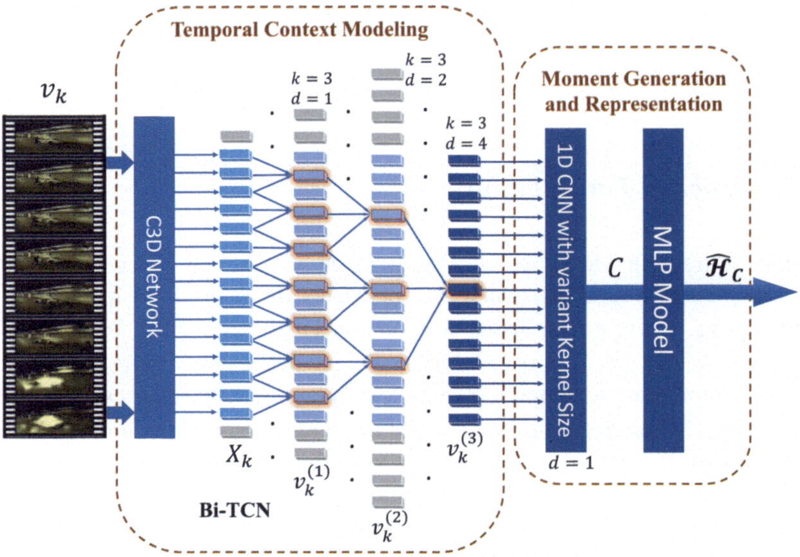

Fig. 7.3 The structure of our proposed VEN module. It first utilizes the C3D model to obtain local features and then incorporates the Bi-TCN model to learn the local features by capturing corresponding pre-context and post-context information. A series of 1-D regular convolutions are subsequently applied to generate moment candidates, and an MLP model is used to obtain feature representations for all the moment candidates

Recently, Temporal Convolutional Networks (TCNs) [2] have emerged as a new strategy for context modeling; however, TCNs focus solely on capturing the relationship between the current local feature and its preceding context, ignoring post-context information. To address this limitation, we design a novel Bi-TCN network capable of capturing long-term contextual dependencies from both pre-context and post-context information for each local feature $\mathbf{x}_{k,i}$, thereby effectively enhancing the contextual representation. Figure 7.3 shows an example of a Bi-TCN with three layers.[2] After the three-layer 1-D dilated convolution process, each element of $\mathbf{v}_k^{(3)}$ integrates contextual information from both directions, resulting in a more comprehensive feature representation. Generally, when the input $\mathbf{X}_k$ is processed through a Bi-TCN with $E-1$ layers, the output is given by:

[2] The dilation factors d for the two hidden layers and the output layer are set to 1, 2, and 4, respectively. The kernel size for all filters is 3. Zero padding of lengths (2, 4, 8) is applied at different layers to ensure that the output feature sequence matches the input length.

$$\begin{cases} \mathbf{v}_k^{(1)} = \theta_1(\mathbf{X}_k, \delta^1, d^1), \\ \vdots \\ \mathbf{v}_k^{(e)} = \theta_e(\mathbf{v}_k^{(e-1)}, \delta^e, d^e), \\ \vdots \\ \mathbf{v}_k^{(E-1)} = \theta_{E-1}(\mathbf{v}_k^{E-2}, \delta^{E-1}, d^{E-1}), \end{cases} \tag{7.1}$$

where θ_e represents the 1-D dilated convolution of the e-th layer, and d^e and δ^e denote the dilation factor and filter kernel size, respectively.

7.3.2.2 Moment Generation and Representation

After obtaining long-term dependencies for each local feature, a series of 1-D regular convolutions (i.e., $d = 1$) with varying kernel sizes δ^* are applied to $\mathbf{v}_k^{(E-1)}$ to generate the moment candidates,[3] as follows:

$$C_k = \{c_{k,1}, \ldots, c_{k,p}, \ldots, c_{k,N_k}\} = \theta_E\left(\mathbf{v}_k^{(E-1)}, \delta^*, d = 1\right), \tag{7.2}$$

where θE represents the 1-D regular convolution, $c_{k,p}$ is the p-th moment candidate, and N_k is the total number of moment candidates.

We then utilize a Multi-Layer Perceptron (MLP) network[4] to compute the feature representations for all the moment candidates in C_k:

$$\begin{cases} \mathcal{H}_{C_k}^1 = \sigma_c^1(\mathbf{W}_c^1 C_k + \mathbf{b}_c^1), \\ \vdots \\ \mathcal{H}_{C_k}^t = \sigma_c^t(\mathbf{W}_c^t \mathcal{H}_{C_k}^{t-1} + \mathbf{b}_c^t), \\ \vdots \\ \mathcal{H}_{C_k}^T = \sigma_c^T(\mathbf{W}_c^T \mathcal{H}_{C_k}^{T-1} + \mathbf{b}_c^T), \end{cases} \tag{7.3}$$

where $\mathbf{W}_c^t$, $\mathbf{b}_c^t$, and $\mathcal{H}_{C_k}^t$ represent the weight matrix, bias vector, and output sequence of the t-th hidden layer, respectively. The activation function σ_c^t used is the Randomized Leaky Rectified Linear Units (RReLU) [31], and the final representations for the moment candidates are given by $\widehat{\mathcal{H}}_{C_k} = \mathcal{H}_{C_k}^T = [\mathbf{h}_{c_{k,1}}, \ldots, \mathbf{h}_{c_{k,p}}, \ldots, \mathbf{h}_{c_{k,N_k}}] \in \mathbb{R}^{N_k \times L}$, where L is the dimensionality of each candidate representation.

---

[3] In this work, δ^* ranges from 1 to Δ, where Δ is a predefined maximum kernel size for candidate generation.

[4] In our experiments, the number of MLP layers is set to two, i.e., $T = 2$.

7.3.3 The QEN Module

To effectively learn query representations, most existing methods rely on Bi-LSTM with word embeddings as input to generate sentence-level representations. However, Bi-LSTM is limited in its ability to capture long-term semantic dependencies, particularly for relatively long queries, which results in suboptimal feature representations. In contrast, the recent advances in language representation, particularly the Bidirectional Encoder Representations from Transformers (BERT) [5], have demonstrated impressive performance in natural language processing. BERT can deeply understand the semantics of a given sentence by leveraging multi-layer bidirectional representations. Therefore, we adopt BERT as the backbone for query encoding in our Query Encoding Network (QEN). Moreover, the pre-trained BERT model is incorporated into QEN and fine-tuned to better capture the semantic details of the queries in our specific task.

As shown in Fig. 7.4, the queries in $\mathbf{Q}_k$ are processed by BERT to output their respective feature representations, denoted as $\widetilde{\mathcal{Q}}_k = \{\widetilde{\mathbf{s}}_{k,1}, \ldots, \widetilde{\mathbf{s}}_{k,j}, \ldots, \widetilde{\mathbf{s}}_{k,M_k}\}$, where $\widetilde{\mathbf{s}}_{k,j}$ represents the feature embedding of the j-th query related to video v_k.

Subsequently, QEN applies a three-layer Multi-Layer Perceptron (MLP) to refine the embeddings of all queries:

$$\begin{cases} \mathcal{H}_{Q_k}^{1} = \sigma_s^1(\mathbf{W}_s^1 \widetilde{\mathcal{Q}}_k + \mathbf{b}_s^1), \\ \quad \vdots \\ \mathcal{H}_{Q_k}^{t} = \sigma_s^t(\mathbf{W}_s^t \mathcal{H}_{Q_k}^{t-1} + \mathbf{b}_s^t), \\ \quad \vdots \\ \mathcal{H}_{Q_k}^{T} = \sigma_s^T(\mathbf{W}_s^T \mathcal{H}_{Q_k}^{T-1} + \mathbf{b}_s^T), \end{cases} \tag{7.4}$$

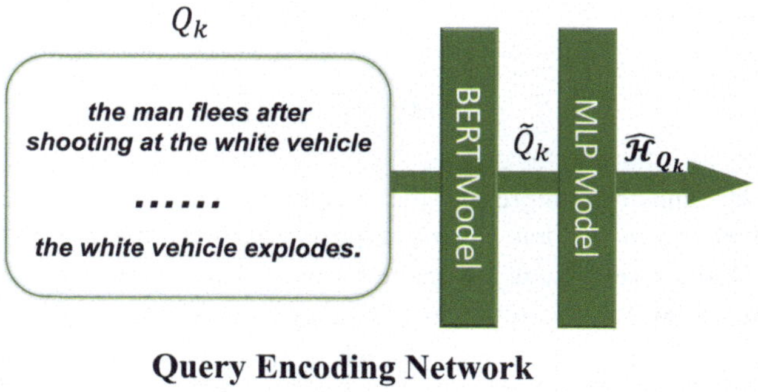

Fig. 7.4 Pipeline of the proposed QEN module. It first utilizes BERT to perform feature encoding for each query in $\mathcal{Q}$. Then, an MLP model is applied to obtain the final representations $\widehat{\mathcal{H}}_Q$

where $\mathbf{W}_s^t$, $\mathbf{b}_s^t$, and $\mathcal{H}_{Q_k}^t$ represent the weight matrix, bias vector, and output sequence of the t-th hidden layer, respectively. The activation function σ_s^t used is the Randomized Leaky Rectified Linear Units (RReLU) [31]. Finally, the output $\widehat{\mathcal{H}}_{Q_k} = \mathcal{H}_{Q_k}^T = [\mathbf{h}_{s_{k,1}}, \ldots, \mathbf{h}_{s_{k,j}}, \ldots, \mathbf{h}_{s_{k,M_k}}] \in \mathbb{R}^{M_k \times L}$ denotes the final representations of the queries, where L represents the dimensionality of each query's representation.

7.3.4 Cross-Modal Hashing

After obtaining the moment representations $\widehat{\mathcal{H}}_{C_k}$, we utilize the element-wise sign function $sgn(\cdot)$ to generate the final hash codes for C_k as follows:

$$\mathcal{B}_{C_k} = sgn(\widehat{\mathcal{H}}_{C_k}) = [\mathbf{b}_{c_{k,1}}, \ldots, \mathbf{b}_{c_{k,p}}, \ldots, \mathbf{b}_{c_{k,N_k}}], \tag{7.5}$$

where $\mathbf{b}_{c_{k,p}} \in \{-1, +1\}^L$ represents the final hash code of the p-th moment candidate with a length of L.

Similarly, the element-wise sign function $sgn(\cdot)$ is applied to transform $\widehat{\mathcal{H}}_{Q_k}$ into the final hash codes for the query set $\widetilde{Q}$:

$$\mathcal{B}_{Q_k} = sgn(\widehat{\mathcal{H}}_{Q_k}) = [\mathbf{b}_{s_{k,1}}, \ldots, \mathbf{b}_{s_{k,j}}, \ldots, \mathbf{b}_{s_{k,M_k}}], \tag{7.6}$$

where $\mathbf{b}_{s_{k,j}} \in \{-1, +1\}^L$ denotes the final hash code of the j-th query, with a length of L.

To ensure that each "moment-query" pair in the Hamming space retains the intrinsic similarity from the original real-valued feature space, we introduce a loss function Γ_1 for semantic similarity preservation:

$$\Gamma_1 = \sum_k \left(\left\| \widehat{\mathcal{H}}_{C_k}^T \widehat{\mathcal{H}}_{Q_k} - L\mathcal{M} \right\|_F^2 \right), \tag{7.7}$$

where $\|\cdot\|_F$ represents the Frobenius norm, L denotes the length of the hash codes, and $\mathcal{M}$ is the cross-modal similarity matrix, designed to ensure that the similarity in the Hamming space aligns with that in the original feature space.

As in conventional cross-modal similarity approaches [12, 26], a straightforward strategy is to use 0 or 1 to represent the similarity of each "moment-query" pair, which can be done in two steps:

1. We use the general evaluation metric Intersection over Union (IoU) [9] to measure the semantic similarities of the $N_k \times M_k$ "moment-query" pairs. Specifically, the IoU between two moments is calculated as follows:

$$IoU_{s_{k,j}}^{c_{k,p}} = \frac{\min\left(t_e^j, \tau_e^p\right) - \max\left(t_s^j, \tau_s^p\right)}{\max\left(t_e^j, \tau_e^p\right) - \min\left(t_s^j, \tau_s^p\right)}, \tag{7.8}$$

where τ_s^p and τ_e^p denote the start and end times of moment $c_{k,p}$, and t_s^j and t_e^j denote the start and end times of the target moment described by query $s_{k,j}$.

2. Each $IoU_{s_{k,j}}^{c_{k,p}}$ is converted to $\mathcal{N}_{p,j} \in \{0, 1\}$ based on a predefined threshold λ. Specifically, if $IoU_{s_{k,j}}^{c_{k,p}} \geq \lambda$, $\mathcal{N}_{p,j}$ is set to 1; otherwise, $\mathcal{N}_{p,j}$ is set to 0. The values of $\mathcal{N}_{p,j}$ are then organized to form the similarity matrix $\mathcal{T} \in \{0, 1\}^{N_k \times M_k}$.

However, using this simplistic (1-or-0) approach cannot accurately capture the complex intrinsic semantic similarities between different "moment-query" pairs. As depicted in Fig. 7.5, both $IoU_{s_{k,1}}^{c_{k,1}}$ and $IoU_{s_{k,1}}^{c_{k,2}}$ are greater than the threshold $\lambda = 0.5$, and hence $\mathcal{N}_{1,1}$ and $\mathcal{N}_{2,1}$ are both set to 1, implying that $c_{k,1}$ and $c_{k,2}$ have identical semantic similarities with $s_{k,1}$. Nonetheless, there are clear differences between $c_{k,1}$ and $c_{k,2}$. Consequently, this simplistic $\mathcal{T}$ can lead to information loss and suboptimal retrieval performance.

To address this limitation, we propose a novel approach to constructing the cross-modal similarity matrix $\mathcal{M}$, as shown in Fig. 7.5. In this approach, each element of $\mathcal{M}$ is directly set to the IoU value, accurately reflecting the similarity of each "moment-query" pair. For instance, the (p, j)-th entry is given by $\mathcal{M}_{p,j} = IoU_{s_{k,j}}^{c_{k,p}}$ $(0 \leq \mathcal{M}_{p,j} \leq 1)$. This soft-value similarity matrix $\mathcal{M}$ enables CMHN to effectively retain the intrinsic similarity of each "moment-query" pair within the Hamming space.

Apart from the semantic preservation regularization, we further regularize the binarization differences between $(\mathbf{h}_{c_{k,p}}, \mathbf{b}_{c_{k,p}})$ and $(\mathbf{h}_{s_{k,j}}, \mathbf{b}_{s_{k,j}})$ to obtain optimal

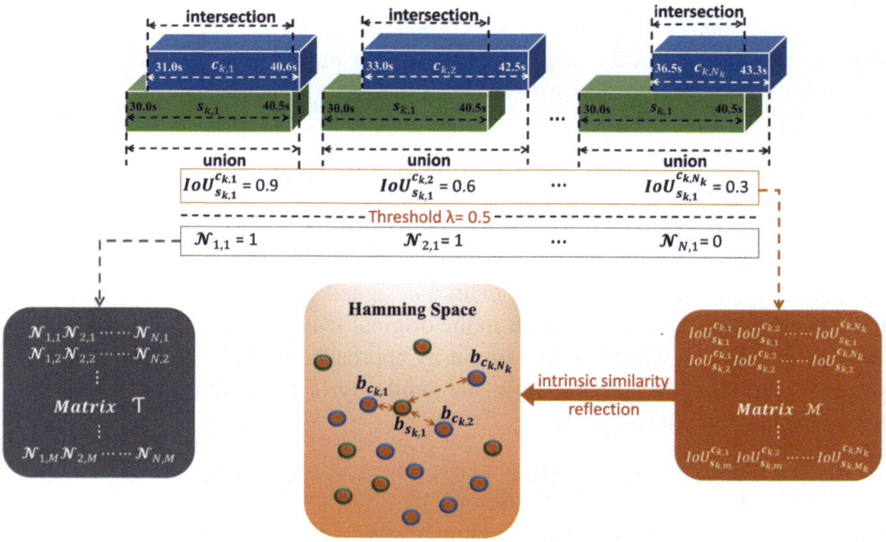

Fig. 7.5 Two types of similarity matrices, $\mathcal{M}$ and $\mathcal{T}$. Compared with the binary similarity metric (0 or 1) used in $\mathcal{T}$, $\mathcal{M}$ is built based on Intersection over Union (IoU) values to better capture the intrinsic similarity of each "moment-query" pair in the Hamming space

continuous surrogates of the binary hash codes:

$$\Gamma_2 = \sum_k \left(\left\| \widehat{\mathcal{H}}_{C_k} - \mathcal{B}_{C_k} \right\|_F^2 + \left\| \widehat{\mathcal{H}}_{Q_k} - \mathcal{B}_{Q_k} \right\|_F^2 \right). \tag{7.9}$$

Furthermore, to balance the learned hash codes and maximize the information conveyed by each bit [26], the representations $\widehat{\mathcal{H}}_{C_k}$ and $\widehat{\mathcal{H}}_{Q_k}$ are regularized as follows:

$$\Gamma_3 = \sum_k \left(\left\| \widehat{\mathcal{H}}_{C_k} \mathcal{U} \right\|_F^2 + \left\| \widehat{\mathcal{H}}_{Q_k} \mathcal{Z} \right\|_F^2 \right), \tag{7.10}$$

where $\mathcal{U} \in \mathbb{R}^{L \times N_k}$ and $\mathcal{Z} \in \mathbb{R}^{L \times M_k}$ are matrices with all elements equal to one.

The final objective function of CMHN is a combination of the three loss functions (Γ_1, Γ_2, and Γ_3):

$$\Psi = \Gamma_1 + \alpha\Gamma_2 + \beta\Gamma_3, \tag{7.11}$$

where α and β are non-negative trade-off parameters.

7.3.5 Inference

As depicted in Fig. 7.6, for an untrimmed video v and its corresponding query set Q, our CMHN first generates the feature representations for both the moment candidates and the queries, denoted by $\widehat{\mathcal{H}}_C$ and $\widehat{\mathcal{H}}_Q$, respectively. These representations are subsequently projected into a shared Hamming space, resulting in the corresponding hash codes $\mathcal{B}_C$ and $\mathcal{B}_Q$.

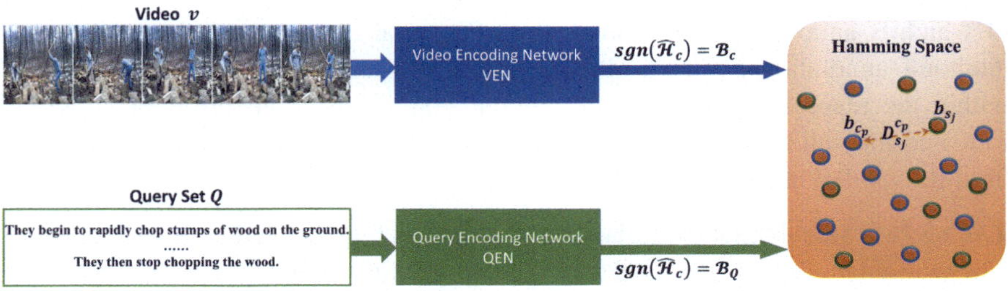

Fig. 7.6 The inference process of our moment localization model, CMHN. After model training, given the video v and its related queries Q, CMHN generates the corresponding hash codes ($\mathcal{B}_C$, $\mathcal{B}_Q$) for different modalities and performs moment retrieval based on the Hamming distance

Once the hash codes are obtained, the similarity between each moment candidate and query is calculated using the Hamming distance:

$$D_{s_j}^{c_p} = \frac{1}{2}(L - \mathbf{b}_{c_p}^T \mathbf{b}_{s_j}), \tag{7.12}$$

where $\mathbf{b}_{c_p}$ and $\mathbf{b}_{s_j}$ represent the hash codes of the p-th moment candidate and the j-th query, respectively. The lower the value of $D_{s_j}^{c_p}$, the greater the similarity between c_p and s_j.

Finally, based on the calculated similarity values for all "moment-query" pairs using Eq. (7.12), the target moment for each query can be efficiently localized by ranking the similarity scores.

7.4 Experiments

To comprehensively validate the effectiveness of the proposed model, we conducted extensive experiments, including comparisons of CMHN with several state-of-the-art methods, analyses of the contribution of each component of our model, as well as experiments to evaluate its efficiency and scalability.

7.4.1 Experimental Settings

7.4.1.1 Datasets

In this chapter, we used two benchmark datasets, namely ActivityNet Captions and TACoS, to evaluate the performance of our proposed model.

ActivityNet Captions [23] This dataset contains 20,000 untrimmed videos, each of which is annotated with an average of 3 natural language queries, along with temporal annotations. Specifically, each query corresponds to a target moment within the respective video. To ensure fair comparison with prior methods, we followed the validation split introduced by Zhu et al. [37], where the val_1 set is used for validation and the val_2 set for testing.

TACoS [6] This dataset comprises 127 videos, each associated with an average of 148 natural language queries and corresponding temporal annotations. Compared to ActivityNet Captions, the videos in TACoS have a longer average duration (ADV), while the average duration of moments (ADM) is relatively shorter.

Detailed statistics for both datasets are summarized in Table 7.1.

Table 7.1 Data statistics of ActivityNet Captions and TACoS datasets, including the number of videos (#Videos), the number of queries (#Queries), average video duration (ADV) in seconds, the average duration of moments (ADM) in seconds, and the average length of queries (ALQ)

	ActivityNet captions dataset					TACoS dataset				
	#Video	#Query	ADV	ADM	ALQ	#Video	#Query	ADV	ADM	ALQ
Training	10,009	37,421	117.30	35.45	13.48	75	10,146	224.16	5.70	8.69
Validation	4917	17,505	118.23	37.73	13.58	27	4589	387.46	6.23	9.12
Testing	4885	17,031	118.21	40.25	12.02	25	4083	367.70	6.96	9.00
All	19,811	71,957	117.74	37.14	13.16	127	18,818	296.21	6.10	8.86

7.4.1.2 Evaluation Metrics

To comprehensively evaluate our model and compare it with baseline methods, we used the "R@n, IoU=m" metric introduced by Hu et al. [9] as the primary evaluation metric for localization accuracy. In the remainder of this paper, we use $R(n, m)$ as shorthand for "R@n, IoU=m".

Additionally, we employed total runtime (TRT) and average runtime (ART) as metrics to assess the efficiency of the models. Specifically, TRT refers to the total time required to complete moment localization for all queries, while ART represents the average time taken to perform moment localization for each query. The ART metric is defined as follows:

$$ART = \frac{TRT}{N_q}, \tag{7.13}$$

where N_q is the total number of queries.

7.4.1.3 Implementation Details

For each video in the datasets, we considered 16 consecutive frames as a unit, with an overlap of 8 frames between adjacent units. These units were then processed by a pre-trained C3D network [27] to produce a 4096-dimensional feature vector for each unit. Specifically, for the ActivityNet Captions dataset, the feature dimensions were further reduced from 4096 to 500 using Principal Component Analysis (PCA) [23]. Consequently, the 500-dimensional and 4096-dimensional features were used as local features for ActivityNet Captions and TACoS, respectively.

All parameters of the VEN and QEN were initialized randomly. We used the Adam optimizer [13] to minimize the multi-task loss function. A grid search strategy was employed to determine the optimal values for the hyperparameters α and β in Eq. (7.11). In this work, we set $\alpha = \beta = 1$ for subsequent experiments.

During training, each batch for the ActivityNet Captions dataset contained an average of 1648 "moment-query" pairs, and the learning rate was set to 0.001. For the TACoS dataset, each batch contained an average of 4865 "moment-query" pairs, with a learning rate of 0.0001. The maximum number of epochs was empirically set to 500 to ensure convergence. All experiments were conducted on a computer running Ubuntu 16.04.6 LTS, equipped with an Intel Xeon CPU E5-2620, 128 GB of RAM, and an NVIDIA TITAN Xp GPU.

7.4.1.4 Hash Code Length Setting

The length of the hash code, denoted as L, is a critical hyperparameter of the proposed CMHN model. Therefore, determining the optimal value for L is essential before conducting the formal performance evaluations. We studied the impact of different values of L on localization accuracy and efficiency. The experimental results on ActivityNet Captions and TACoS are summarized in Table 7.2.

The results indicate that as the hash code length L increases, the localization accuracy of CMHN improves, but the corresponding runtime also increases. This observation suggests that longer hash codes can retain more information and improve accuracy, but they also degrade localization efficiency. To balance accuracy and efficiency, we set the hash code length to 256 for ActivityNet Captions and 128 for TACoS.

7.4.2 Performance Comparison

To evaluate the effectiveness of our proposed CMHN, we compared it with six state-of-the-art baseline methods: MCN [1], CTRL [6], ACRN [18], ABLR [35], QSPN [32], and CMIN [37]. Tables 7.3 and 7.4 present the overall localization accuracy of our CMHN compared to the baselines on the ActivityNet Captions and TACoS datasets, respectively. For both datasets, we used the unified evaluation metric "R@n, IoU=m" with $n \in \{1, 5\}$ and $m \in \{0.3, 0.5, 0.7\}$.

Based on the results presented in Tables 7.3 and 7.4, we draw the following observations:

- Both MCN and CTRL exhibit lower performance compared to other models, primarily because they incorporate contextual information either by roughly fusing the entire video features or by extending moment boundaries over a limited scale. Specifically, MCN fuses a large amount of contextual information, which can introduce noise and reduce the quality of the context representation. CTRL, on the other hand, considers only a limited extension of pre- and post-moments, which fails to model longer-term dependencies effectively.
- ACRN, QSPN, and ABLR outperform MCN and CTRL, indicating that the use of attention mechanisms helps to highlight critical modality information and enhances feature representation. Notably, ABLR directly regresses temporal locations based on visual-textual co-attention weights/features, which allows it to generate only one prediction per query.

Table 7.2 Performance of CMHN with different hash code lengths (L) on the ActivityNet Captions and TACoS datasets. Specifically, we reported results for "R@n, IoU=m" with $n \in 1, 5$ and $m \in 0.5, 0.7$

	ActivityNet captions					TACoS				
	R@1	R@1	R@5	R@5		R@1	R@1	R@5	R@5	
L	IoU=0.7	IoU=0.5	IoU=0.7	IoU=0.5	TRT(s)	IoU=0.7	IoU=0.5	IoU=0.7	IoU=0.5	TRT(s)
32	17.66%	38.42%	46.89%	70.54%	127.70	8.55%	19.94%	20.94%	32.87%	3.61
64	19.72%	40.59%	48.86%	71.92%	128.29	12.98%	23.17%	25.32%	34.63%	3.62
128	22.12%	42.45%	51.27%	73.08%	129.29	18.44%	25.58%	28.24%	35.23%	3.65
256	24.02%	43.47%	53.16%	73.42%	130.17	19.98%	26.08%	28.53%	36.69%	3.73
512	24.12%	43.87%	53.26%	73.52%	131.03	20.72%	26.23%	29.37%	36.72%	3.84

Table 7.3 Localization accuracy comparison between our proposed model and several state-of-the-art baselines on the ActivityNet Captions dataset. (p-value*: p-value over $R(1, 0.5)$)

Method	R@1 IoU=0.7	R@1 IoU=0.5	R@1 IoU=0.3	R@5 IoU=0.7	R@5 IoU=0.5	R@5 IoU=0.3	p-value*
MCN	6.43%	21.36%	39.35%	29.70%	53.23%	68.12%	4.16E-17
CTRL	10.34%	29.01%	47.43%	37.54%	59.17%	75.32%	1.83E-15
ACRN	11.25%	31.67%	49.70%	38.57%	60.34%	76.50%	1.11E-14
QSPN	13.43%	33.26%	52.13%	40.78%	62.39%	77.72%	8.38E-14
ABLR	15.71%	36.79%	55.67%	–	–	–	1.68E-12
CMIN	23.88%	43.40%	**63.61%**	50.73%	67.95%	80.54%	4.68E-03
CMHN	**24.02%**	**43.47%**	62.49%	**53.16%**	**73.42%**	**85.37%**	–

The best experimental results are highlighted in bold

Table 7.4 Localization accuracy comparison between our proposed model and several state-of-the-art baselines on the TACoS dataset. (p-value*: p-value over $R(1, 0.5)$)

Method	R@1 IoU=0.7	R@1 IoU=0.5	R@1 IoU=0.3	R@5 IoU=0.7	R@5 IoU=0.5	R@5 IoU=0.3	p-value*
MCN	1.00%	1.25%	1.64%	1.01%	1.25%	2.03%	6.11E-14
CTRL	6.98%	13.08%	18.17%	15.60%	26.13%	35.97%	2.40E-11
ACRN	8.28%	14.11%	20.06%	15.43%	26.30%	38.11%	5.18E-11
QSPN	7.56%	15.23%	20.15%	14.35%	25.30%	36.72%	1.30E-10
ABLR	6.13%	9.30%	18.90%	–	–	–	2.25E-12
CMIN	7.64%	18.05%	24.64%	11.81%	27.02%	38.46%	2.20E-09
CMHN	**18.44%**	**25.58%**	**30.04%**	**28.24%**	**35.23%**	**44.05%**	–

The best experimental results are highlighted in bold

- CMIN achieves relatively higher accuracy compared to other baselines, verifying the importance of modeling long-range semantic dependencies for effective moment localization.
- CMHN achieves the highest localization accuracy across all scenarios, significantly outperforming state-of-the-art baselines, particularly on the TACoS dataset. This demonstrates the effectiveness of our proposed model. Furthermore, we conducted significance tests between CMHN and each baseline using R(1,0.5) based on 20-round experimental results. The p-values were all smaller than 0.05, indicating that the improvement of CMHN over the baselines is statistically significant.

Interestingly, the localization accuracy of all models on TACoS is considerably lower than that on ActivityNet Captions. The primary reasons are as follows: (1) The longer ADV and shorter ADM in TACoS make effective moment representation more challenging. (2) TACoS has more queries and shorter ALQ, which increases the difficulty of distinguishing similar moments associated with different queries. In general, moment localization on TACoS is more challenging compared to ActivityNet Captions.

7.4.3 Ablation Study

We conducted an ablation study to evaluate the contributions of the video encoding, query encoding, and cross-modal hashing components in our model. Specifically, we created several ablation models by removing or replacing one component at a time:

- **CMHN-V**: We replaced the proposed Bi-TCN in VEN with traditional Bi-LSTM for video encoding.
- **CMHN-T**: We replaced the proposed Bi-TCN in VEN with a traditional TCN for video encoding.
- **CMHN-S**: We replaced BERT in QEN with GloVe word embeddings [22] and Bi-LSTM for query encoding.
- **CMHN-H**: We eliminated the proposed cross-modal similarity matrix (CSM) and used a hard similarity matrix $\mathcal{T}$ (threshold $\lambda = 0.5$) to evaluate "moment-query" pair similarities.

The ablation study was performed on the ActivityNet Captions and TACoS datasets, with results summarized in Fig. 7.7. From these results, we make the following observations:

- CMHN-H exhibits the poorest performance on both datasets, particularly at higher IoU thresholds (e.g., IoU $\in \{0.6, 0.7, 0.8, 0.9\}$). This highlights that a hard similarity representation (1-or-0) fails to capture the nuanced intrinsic semantic similarities between "moment-query" pairs, leading to suboptimal localization accuracy. The significant improvement achieved by CMHN underscores the effectiveness of the proposed soft similarity matrix for cross-modal hashing.
- CMHN-S, CMHN-V, and CMHN-T underperform compared to CMHN, indicating that: (1) Word2Vec-based query encoding lacks the semantic depth and contextual understanding provided by BERT. (2) Bi-LSTM struggles with tem-

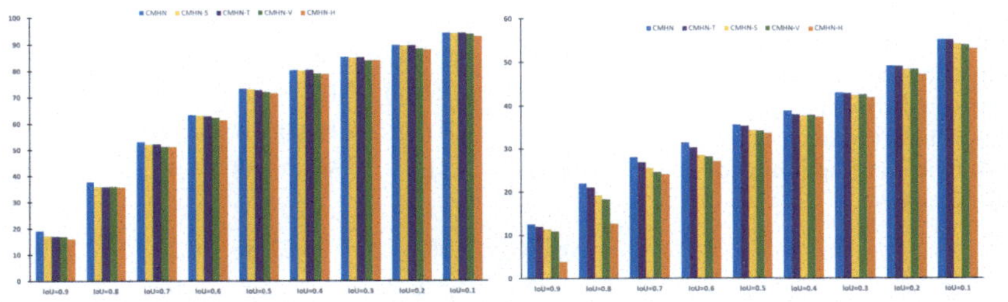

(a) R@5 w.r.t IoU on ActivityNet Captions (b) R@5 w.r.t IoU on TACoS

Fig. 7.7 Localization accuracy comparison between CMHN and its variants on the ActivityNet Captions and TACoS datasets. (**a**) Experimental results for R@5 with respect to IoU $\in 0.1, \ldots, 0.9$ on the ActivityNet Captions dataset. (**b**) Experimental results for R@5 with respect to IoU $\in 0.1, \ldots, 0.9$ on the TACoS dataset poral context modeling in relatively long videos. (3) TCN fails to simultaneously capture pre-context and post-context information, resulting in less effective video encoding compared to Bi-TCN.

- CMHN outperforms all variants across both datasets, demonstrating the importance of Bi-TCN, BERT, and the proposed CSM in achieving superior moment localization.

7.4.4 Efficiency Comparison

To evaluate the efficiency of our proposed CMHN, we compared its runtime performance with state-of-the-art baseline methods on the ActivityNet Captions and TACoS datasets. Table 7.5 reports the ART and TRT for each method. From these results, the following insights emerge:

- ABLR demonstrates smaller ART and TRT compared to the two-stage "scan and localize" architectures such as MCN, CTRL, and ACRN. This is because these models generate densely overlapping moment candidates using the sliding window approach for cross-modal localization. In contrast, ABLR avoids redundant computations by evenly dividing the original video into moment candidates and performing temporal coordinate regression directly.
- CMIN achieves significantly lower ART and TRT than other baselines. This is primarily due to its preprocessing steps: reducing the original video feature dimensions from 4096 to 500 using PCA, and further downsampling the feature sequences to 200 dimensions. These operations accelerate video encoding while improving overall efficiency.
- CMHN outperforms all baselines in efficiency. Particularly on the TACoS dataset, CMHN achieves an order-of-magnitude higher efficiency compared to CMIN. These results highlight the superior efficiency of the proposed model.

Table 7.5 Running time comparison between our proposed CMHN and state-of-the-art baselines on the ActivityNet Captions and TACoS datasets

	ActivityNet captions		TACoS	
Method	ART (s)	TRT (s)	ART (s)	TRT (s)
MCN	6.97	118,706.07	15.68	64,021.44
CTRL	3.41	58,135.38	5.52	26,515.75
ACRN	4.42	75,235.08	6.94	33,457.43
ABLR	0.06	1,021.86	0.23	958.50
CMIN	0.72E-2	123.18	1.03E-2	41.93
CMHN	0.76E-2	130.17	**0.90E-3**	**3.65**

The best experimental results are highlighted in bold

7.4.5 Scalability Comparison

The workload of cross-modal moment localization varies significantly between the two datasets. For instance, the average number of moments to be localized per video in ActivityNet Captions is 3, while in TACoS, it is 148. Since CMIN achieves slightly better ART than CMHN on ActivityNet Captions (7.2 ms vs. 7.6 ms), we further analyzed the scalability of both models on this dataset.

Both models were trained offline using the same training set, and their runtime performance was evaluated as the number of online queries increased. The results are shown in Fig. 7.8. From these results, the following insights emerge:

- TRT Comparison (Fig. 7.8a): Both models exhibit increasing TRT as the number of queries rises. However, the growth rate of CMHN is significantly lower than

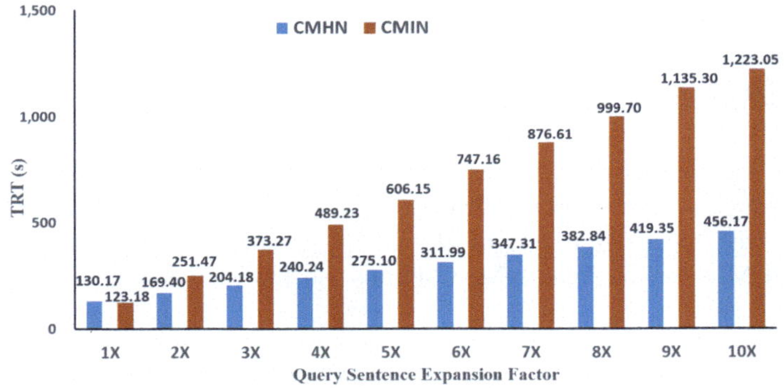

(a) TRT on ActivityNet Captions

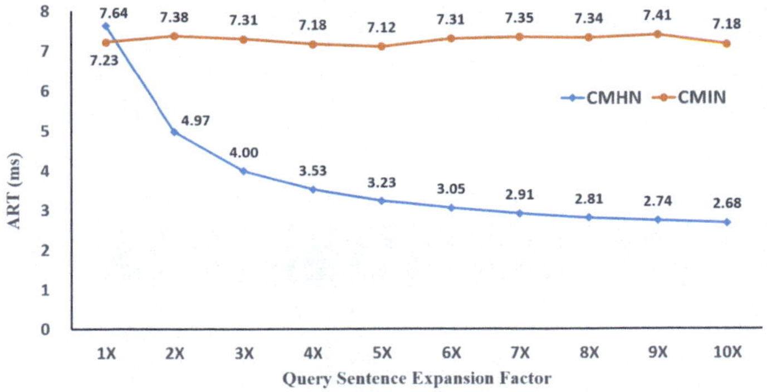

(b) ART on ActivityNet Captions

Fig. 7.8 Scalability comparison between CMHN and CMIN on the ActivityNet Captions dataset with different numbers of queries. "2X" indicates that the queries in the testing set are doubled, i.e., $34,062 = 2 \times 17,031$. (**a**) TRT results for {1X, 2X, ..., 10X} queries. (**b**) ART results for {1X, 2X, ..., 10X} queries that of CMIN. This is because CMIN's attentive aggregation module heavily depends on video information. For each new query, CMIN must re-execute the video encoder and the aggregation module to extract enhanced representations, leading to a sharp rise in TRT. In contrast, CMHN's VEN and QEN modules are independent, allowing for offline learning of hash codes for any new videos and queries, thus improving localization efficiency. This independence makes CMHN less affected by an increasing number of queries.

- ART Comparison (Fig. 7.8b): CMIN's ART remains relatively constant, fluctuating around an average of 7.28 ms, indicating a consistent average runtime for localizing one moment per query. On the other hand, CMHN's ART continuously decreases as the number of queries increases. This demonstrates that the TRT growth rate of CMHN is much smaller than the rate at which queries increase, as defined by Eq. (7.13). These results confirm that CMHN exhibits superior scalability compared to state-of-the-art baselines.

7.4.6 *Qualitative Results*

To further validate the effectiveness of our proposed CMHN model, we presented several examples of moment localization for different queries. Specifically, Fig. 7.9 illustrates the localization results of CMHN compared to the best-performing baseline, CMIN, on the ActivityNet Captions and TACoS datasets.

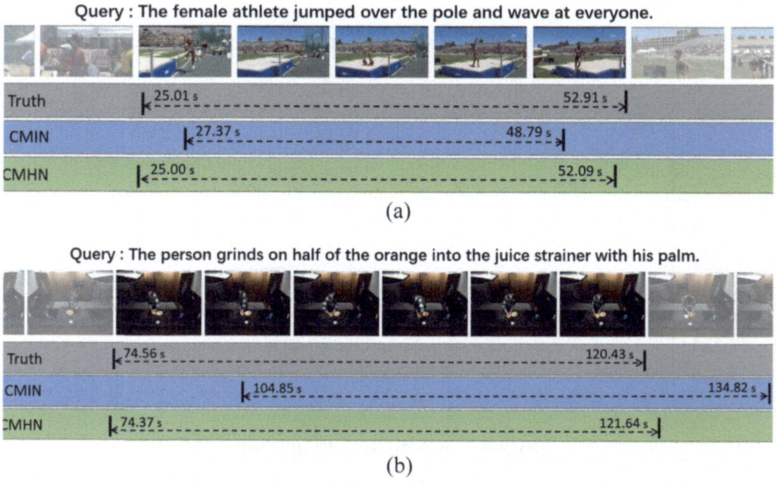

Fig. 7.9 Moment localization results on the ActivityNet Captions and TACoS datasets. Both figures present R@1 results across various IoU thresholds. (**a**) Moment localization result on ActivityNet Captions. (**b**) Moment localization result on TACoS

From Fig. 7.9, it is evident that CMHN achieves more accurate results, as the localized moments returned by our model exhibit higher IoU with the ground truth moments compared to those returned by CMIN. The superior performance of CMHN can be attributed to two key factors: (1) The Bi-TCN-based CMHN effectively captures long-term contextual semantic dependencies, outperforming the Bi-GRU-based CMIN in the modeling video context. (2) By leveraging an IoU-based cross-modal similarity matrix, CMHN better reflects the semantic similarity of each "moment-query" pair. This facilitates more effective cross-modal hash code learning, enabling CMHN to deliver more precise moment localization results.

7.5 Summary

In this chapter, we propose an end-to-end deep cross-modal hashing network to address the task of video moment localization. Specifically, we incorporate a bidirectional temporal convolutional network into the video encoder to effectively model temporal contextual structures across multiple scales, enabling the simultaneous generation of moment candidates and their enhanced representations. Complementing this, we design an independent query encoder to accurately interpret user requirements. To bridge the modalities, we introduce a cross-modal hashing module for learning compact hash codes. Utilizing these hash codes, the relevance scores for each "moment-query" pair are efficiently estimated via the Hamming distance. To validate the effectiveness, efficiency, and scalability of CMHN, we conducted extensive experiments on two public benchmark datasets, comparing our model against several state-of-the-art baselines. The results demonstrated the superior performance of CMHN in terms of accuracy and computational efficiency.

References

1. Anne Hendricks, L., Wang, O., Shechtman, E., Sivic, J., Darrell, T., Russell, B.: Localizing moments in video with natural language. In: Proceedings of the IEEE Conference on Computer Vision and Pattern Recognition, pp. 5803–5812 (2017)
2. Bai, S., Kolter, J.Z., Koltun, V.: Convolutional sequence modeling revisited. In: Proceedings of International Conference on Learning Representations, pp. 1–14 (2018)
3. Cui, H., Zhu, L., Li, J., Yang, Y., Nie, L.: Scalable deep hashing for large-scale social image retrieval. IEEE Trans. Image Process. **29**, 1271–1284 (2019)
4. De Boer, M.H., Lu, Y.J., Zhang, H., Schutte, K., Ngo, C.W., Kraaij, W.: Semantic reasoning in zero example video event retrieval. ACM Trans. Multimedia Comput. Commun. Appl. **13**, 1–17 (2017)
5. Devlin, J., Chang, M.W., Lee, K., Toutanova, K.: Bert: pre-training of deep bidirectional transformers for language understanding. In: Proceedings of the Conference of the North American Chapter of the Association for Computational Linguistics: Human Language Technologies, pp. 4171–4186 (2019)

6. Gao, J., Sun, C., Yang, Z., Nevatia, R.: Tall: temporal activity localization via language query. In: Proceedings of the IEEE International Conference on Computer Vision, pp. 5267–5275 (2017)
7. Gao, J., Yang, Z., Sun, C., Chen, K., Nevatia, R.: Turn tap: temporal unit regression network for temporal action proposals. In: Proceedings of the IEEE International Conference on Computer Vision, pp. 3628–3636 (2017)
8. Hao, Y., Mu, T., Goulermas, J.Y., Jiang, J., Hong, R., Wang, M.: Unsupervised t-distributed video hashing and its deep hashing extension. IEEE Trans. Image Process. **26**, 5531–5544 (2017)
9. Hu, R., Xu, H., Rohrbach, M., Feng, J., Saenko, K., Darrell, T.: Natural language object retrieval. In: Proceedings of the IEEE Conference on Computer Vision and Pattern Recognition, pp. 4555–4564 (2016)
10. Huang, H., Jain, V., Mehta, H., Ku, A., Magalhaes, G., Baldridge, J., Ie, E.: Transferable representation learning in vision-and-language navigation. In: Proceedings of the IEEE International Conference on Computer Vision, pp. 7404–7413 (2019)
11. Irie, G., Arai, H., Taniguchi, Y.: Alternating co-quantization for cross-modal hashing. In: IEEE International Conference on Computer Vision, pp. 1886–1894 (2015)
12. Jiang, Q., Li, W.: Deep cross-modal hashing. Comput. Res. Repos. **abs/1602.02255**, 1–12 (2016)
13. John, D., Elad, H., Yoram, S.: Adaptive subgradient methods for online learning and stochastic optimization. J. Mach. Learn. Res. **12**, 2121–2159 (2011)
14. Li, S., Chen, Z., Lu, J., Li, X., Zhou, J.: Neighborhood preserving hashing for scalable video retrieval. In: Proceedings of the IEEE International Conference on Computer Vision, pp. 8212–8221 (2019)
15. Li, Y., Liu, M., Yin, J., Cui, C., Xu, X.S., Nie, L.: Routing micro-videos via a temporal graph-guided recommendation system. In: Proceedings of the 27th ACM International Conference on Multimedia, pp. 1464–1472 (2019)
16. Lin, Z., Ding, G., Han, J., Wang, J.: Cross-view retrieval via probability-based semantics-preserving hashing. IEEE Trans. Cybern. **47**, 4342–4355 (2017)
17. Liu, M., Nie, L., Wang, M., Chen, B.: Towards micro-video understanding by joint sequential-sparse modeling. In: Proceedings of the 25th ACM International Conference on Multimedia, pp. 970–978 (2017)
18. Liu, M., Wang, X., Nie, L., He, X., Chen, B., Chua, T.S.: Attentive moment retrieval in videos. In: Proceedings of the ACM SIGIR Conference on Research and Development in Information Retrieval, pp. 15–24 (2018)
19. Liu, M., Wang, X., Nie, L., Tian, Q., Chen, B., Chua, T.: Cross-modal moment localization in videos. In: Proceedings of the ACM International Conference on Multimedia, pp. 843–851 (2018)
20. Liu, M., Nie, L., Wang, X., Tian, Q., Chen, B.: Online data organizer: micro-video categorization by structure-guided multimodal dictionary learning. IEEE Trans. Image Process. **28**, 1235–1247 (2019)
21. Lu, X., Chen, Y., Li, X.: Hierarchical recurrent neural hashing for image retrieval with hierarchical convolutional features. IEEE Trans. Image Process. **27**, 106–120 (2018)
22. Pennington, J., Socher, R., Manning, C.: Glove: global vectors for word representation. In: Proceedings of the Conference on Empirical Methods in Natural Language Processing, pp. 1532–1543 (2014)
23. Ranjay, K., Kenji, H., Frederic, R., Li, F.F., Juan, C.N.: Dense-captioning events in videos. In: Proceedings of the IEEE International Conference on Computer Vision, pp. 706–715 (2017)
24. Schoeffmann, K., Hopfgartner, F.: Interactive video search. In: Proceedings of the ACM International Conference on Multimedia, pp. 1321–1322 (2015)
25. Song, J., Yang, Y., Yang, Y., Huang, Z., Shen, H.T.: Inter-media hashing for large-scale retrieval from heterogeneous data sources. In: Proceedings of the ACM SIGMOD International Conference on Management of Data, pp. 785–796 (2013)

26. Sun, C., Song, X., Feng, F., Zhao, W.X., Zhang, H., Nie, L.: Supervised hierarchical cross-modal hashing. In: Proceedings of the ACM SIGIR Conference on Research and Development in Information Retrieval, pp. 725–734 (2019)
27. Tran, D., Bourdev, L., Fergus, R., Torresani, L., Paluri, M.: Learning spatiotemporal features with 3d convolutional networks. In: Proceedings of the IEEE International Conference on Computer Vision, pp. 4489–4497 (2015)
28. Vallet, D., Hopfgartner, F., Jose, J.M., Castells, P.: Effects of usage-based feedback on video retrieval: a simulation-based study. ACM Trans. Inf. Syst. **29**, 11–32 (2011)
29. Wang, X., Huang, Q., Celikyilmaz, A., Gao, J., Shen, D., Wang, Y.F., Wang, W.Y., Zhang, L.: Reinforced cross-modal matching and self-supervised imitation learning for vision-language navigation. In: Proceedings of the IEEE Conference on Computer Vision and Pattern Recognition, pp. 6629–6638 (2019)
30. Wei, Y., Wang, X., Guan, W., Nie, L., Lin, Z., Chen, B.: Neural multimodal cooperative learning toward micro-video understanding. IEEE Trans. Image Process. **29**, 1–14 (2020)
31. Xu, B., Wang, N., Chen, T., Li, M.: Empirical evaluation of rectified activations in convolutional network. In: Proceedings of the International Conference on Machine Learning, pp. 1–5 (2015)
32. Xu, H., He, K., Plummer, B.A., Sigal, L., Sclaroff, S., Saenko, K.: Multilevel language and vision integration for text-to-clip retrieval. In: Proceedings of the American Association for Artificial Intelligence, pp. 9062–9069 (2019)
33. Yan, R., Yang, J., Hauptmann, A.G.: Learning query-class dependent weights in automatic video retrieval. In: Proceedings of the ACM International Conference on Multimedia, pp. 548–555 (2004)
34. Yu, Z., Wu, F., Yang, Y., Tian, Q., Luo, J., Zhuang, Y.: Discriminative coupled dictionary hashing for fast cross-media retrieval. In: Proceedings of the International ACM SIGIR Conference on Research and Development in Information Retrieval, pp. 395–404 (2014)
35. Yuan, Y., Mei, T., Zhu, W.: To find where you talk: temporal sentence localization in video with attention based location regression. In: Proceedings of the American Association for Artificial Intelligence, pp. 9159–9166 (2019)
36. Zhou, J., Ding, G., Guo, Y.: Latent semantic sparse hashing for cross-modal similarity search. In: Proceedings of the International ACM SIGIR Conference on Research and Development in Information Retrieval, pp. 415–424 (2014)
37. Zhu, Z., Zhijie, L., Zhou, Z., Zhenxin, X.: Cross-modal interaction networks for query-based moment retrieval in videos. In: Proceedings of the International ACM SIGIR Conference on Research and Development in Information Retrieval, pp. 655–664 (2019)

Chapter 8
Research Frontiers

8.1 Introduction

Thus far, this book has explored the task of video moment localization, which aims to identify specific temporal segments within a video that correspond to a natural language query. We began by highlighting the key challenges of this task, including effective temporal context modeling, cross-modal semantic alignment, efficient candidate pruning, and scalable computation. To address these challenges, we proposed a series of advanced techniques designed to enhance the accuracy and efficiency of video moment localization. First, we introduced the semantic enhanced video moment localization method, emphasizing the importance of understanding both the target moment and its surrounding context. This approach employs a memory attention mechanism to prioritize visual features referenced in the query while incorporating broader contextual information. To improve semantic alignment, we presented the semantic alignment video moment localization method, which dynamically decomposes queries into relevant and irrelevant components. This method ensures precise alignment between textual and visual modalities, particularly for queries containing complex temporal keywords. Addressing computational inefficiency caused by redundant moment candidates, we proposed the semantic pruning video moment localization method. This approach incorporates a cross-modal semantic alignment network and a pruning strategy to reduce unnecessary computations while enhancing localization accuracy. Next, we developed the semantic collaborative video moment localization framework, addressing limitations in integrating intra-modal and inter-modal information. By fostering collaboration between visual and textual modalities, this method achieves a hierarchical semantic understanding, leading to more effective localization. To overcome the dependency on expensive moment-level annotations, we introduced the weakly-supervised video moment localization method, which relies solely on video-level annotations. Using a context-aware multiple-instance learning module, this method bridges the semantic gap between modalities and improves retrieval

M. Liu et al., *Intelligent and Efficient Video Moment Localization*,
https://doi.org/10.1007/978-3-031-87588-5_8

performance. Finally, we presented the efficient hashing-based video moment localization method, designed for scalability and efficiency. By utilizing compact hash codes for fast retrieval and enabling offline computation of video and query representations, this method significantly enhances performance on large-scale video datasets.

The theories and methodologies for video moment localization presented in this book benefit a diverse range of stakeholders, including researchers, developers, and end-users. Researchers and developers can leverage these advanced techniques for semantic alignment, contextual understanding, and efficient cross-modal retrieval as a foundation for building next-generation vision-language systems. Applications include video indexing, summarization, and intelligent surveillance, where precise moment identification within vast video archives is critical. Online platforms and content providers can use these methods to enhance user engagement and satisfaction. Video streaming services can implement personalized search systems to help users locate specific scenes efficiently. Similarly, e-learning platforms can enable students to identify relevant sections of educational videos based on their queries, greatly improving the learning experience. End-users, the primary beneficiaries, gain significantly from improved video interaction. With these techniques, users can effortlessly search for specific moments in untrimmed videos using natural language queries, eliminating the frustration of manually scanning lengthy content. This capability transforms video interaction across entertainment, education, and professional domains.

Despite these advancements, video moment localization remains a dynamic and evolving research area with vast potential for growth. In the next section, we outline promising future research directions and the associated challenges that lie ahead.

8.2 Unsupervised Video Moment Localization

Recent advancements in deep learning and the availability of large-scale annotated datasets have significantly improved the performance of video moment localization models. These datasets typically include temporal annotations (start and end times) paired with natural language queries, enabling supervised training. However, generating such annotations is labor-intensive and costly. Weakly-supervised approaches have emerged as an alternative, requiring only video-level annotations to localize moments. Although these methods reduce the need for precise temporal annotations, annotating natural language queries remains a bottleneck, limiting scalability.

To overcome this challenge, unsupervised video moment localization aims to eliminate the need for paired annotations altogether. This novel paradigm offers a cost-effective and scalable solution for natural language-based video moment localization. In the unsupervised setup, no annotated queries or temporal alignments are available. Recent methods address this by constructing pseudo-supervision through semantic concepts. For instance, PSVL [8] generates pseudo-queries by simulating supervised training, inspired by unsupervised machine translation [1].

Gao and Xu [4] proposed a video-conditioned sentence generator to create pseudo-queries based on mined video concepts. Similarly, SPL [11] leverages large-scale pre-trained models to generate free-form sentences, achieving strong performance, albeit with dependencies on downstream model training.

Beyond pseudo-supervision, approaches such as DSCNet [7] explore video-text alignment by leveraging semantic relationships within video and text datasets, eliminating the need for paired annotations. These methods show promise for reducing annotation dependency while maintaining performance.

Leveraging large-scale vision-language models (VLMs) represents an exciting direction for unsupervised video moment localization. However, fine-tuning such models often compromises their generalizability. A key challenge lies in leveraging the pre-trained video-text alignments of VLMs while minimizing reliance on task-specific datasets. Future research must focus on harnessing the temporal and semantic richness of VLMs to create scalable, generalizable solutions for unsupervised moment localization.

8.3 Long Video Moment Localization

Understanding long-form videos, ranging from minutes to hours, introduces unique challenges for video moment localization. These tasks demand models capable of processing extensive sequences, capturing both local and global temporal structures, and reasoning over diverse temporal contexts. Unlike short clips, long-form videos require the ability to handle dependencies across distant moments while ensuring computational efficiency.

One significant challenge is scalability in temporal context modeling. Current methods often process long videos by dividing them into overlapping temporal windows [3], applying grounding methods within each window. However, this approach fragments temporal context and risks false positives, especially in repetitive or ambiguous content.

Another limitation lies in the disjoint architectures of proposal-based methods [6, 9], which separate clip generation and grounding. This leads to inefficiencies and difficulties in locating short, precise moments within long videos. Robust semantic alignment between queries and video content remains a critical challenge, exacerbated by the inherent ambiguity of natural language.

To address these issues, future approaches should focus on: (1) **Unified Architectures:** End-to-end frameworks integrating clip retrieval and grounding for seamless processing. (2) **Hierarchical Temporal Modeling:** Segmenting videos into granularities for multi-scale analysis, balancing local details and global context. (3) **Dynamic Temporal Sampling:** Using techniques like keyframe selection or reinforcement learning to focus on query-relevant segments, reducing computational overhead.

8.4 Video Corpus Moment Localizaiton

While video moment localization targets single videos, real-world applications often require querying across large collections of untrimmed, unsegmented videos. Video corpus moment localization [5] addresses this need, enabling temporal localization of relevant moments across vast datasets based on natural language queries. This capability is transformative for video search, surveillance, and content curation.

The primary challenge lies in scalability. Retrieving moments across thousands of videos requires balancing computational efficiency and retrieval accuracy. Existing methods rely on exhaustive similarity computations, which are computationally intensive. Future directions include: (1) **Advanced Indexing:** Techniques like compressed representations, hashing, or graph-based retrieval structures for approximate nearest-neighbor searches. (2) **Efficient Retrieval Frameworks:** Combining indexing with pre-trained embeddings to optimize retrieval performance without compromising accuracy.

8.5 Streaming Video Moment Localization

The exponential growth of streaming videos, including surveillance and live-stream content, has underscored the need for advanced algorithms capable of identifying the precise moment a queried event occurs in real-time. Streaming video moment localization addresses this challenge by dynamically evaluating the relevance of a natural language query to an ongoing video stream. For instance, given a query such as "Something is blown to the street by a typhoon," a streaming video moment localization model must estimate, in real-time, whether the current moment corresponds to the start, middle, or end of the event.

Despite its potential, streaming video moment localization presents several key challenges: (1) **Incomplete Information Inference:** Streaming models operate without future frames, relying on past and current data. Forecasting approaches, effective for predefined actions, struggle with natural language queries due to their complexity and variability. (2) **Historical Information Compression:** Processing long temporal sequences in streaming data risks diluting focus on query-relevant content. Effective compression methods must integrate natural language queries to filter out irrelevant details. Future solutions should explore techniques like dynamic memory networks and adaptive compression to balance real-time processing demands with accurate localization.

8.6 Video Grounding Question Answering

Video grounding question answering (Video Grounding QA) builds upon video moment localization by requiring models to both identify relevant video moments and generate visually grounded answers to natural language queries. Unlike traditional video moment localization, which focuses on identifying temporal video segments that correspond to natural language queries, Video Grounding QA requires models to not only locate the relevant video moments but also generate contextually accurate and visually grounded answers. This fusion of temporal understanding and reasoning makes Video Grounding QA a more comprehensive and challenging task in the vision-language domain.

Datasets like NExT-GQA [10] and QAEgo4D [2] bridge video moment localization and question answering, providing annotated temporal segments tied to answers. By leveraging these resources, future research can develop models that excel in multimodal understanding while ensuring interpretability and trustworthiness.

References

1. Artetxe, M., Labaka, G., Agirre, E., Cho, K.: Unsupervised neural machine translation. In: International Conference on Learning Representations, pp. 3636–3647 (2018)
2. Bärmann, L., Waibel, A.: Where did i leave my keys?-episodic-memory-based question answering on egocentric videos. In: Proceedings of the IEEE/CVF Conference on Computer Vision and Pattern Recognition, pp. 1560–1568 (2022)
3. Barrios, W., Soldan, M., Ceballos-Arroyo, A.M., Heilbron, F.C., Ghanem, B.: Localizing moments in long video via multimodal guidance. In: Proceedings of the IEEE/CVF International Conference on Computer Vision, pp. 13667–13678 (2023)
4. Gao, J., Xu, C.: Learning video moment retrieval without a single annotated video. IEEE Trans. Circuits Syst. Video Technol. **32**(3), 1646–1657 (2021)
5. Hou, Z., Ngo, C.W., Chan, W.K.: Conquer: contextual query-aware ranking for video corpus moment retrieval. In: Proceedings of the 29th ACM International Conference on Multimedia, pp. 3900–3908 (2021)
6. Hou, Z., Zhong, W., Ji, L., Gao, D., Yan, K., Chan, W., Ngo, C.W., Shou, M.Z., Duan, N.: Cone: an efficient coarse-to-fine alignment framework for long video temporal grounding. In: Proceedings of the 61st Annual Meeting of the Association for Computational Linguistics (Volume 1: Long Papers), pp. 8013–8028 (2023)
7. Liu, D., Qu, X., Wang, Y., Di, X., Zou, K., Cheng, Y., Xu, Z., Zhou, P.: Unsupervised temporal video grounding with deep semantic clustering. In: Proceedings of the AAAI Conference on Artificial Intelligence, vol. 36, pp. 1683–1691 (2022)
8. Nam, J., Ahn, D., Kang, D., Ha, S.J., Choi, J.: Zero-shot natural language video localization. In: Proceedings of the IEEE/CVF International Conference on Computer Vision, pp. 1470–1479 (2021)
9. Pan, Y., He, X., Gong, B., Lv, Y., Shen, Y., Peng, Y., Zhao, D.: Scanning only once: an end-to-end framework for fast temporal grounding in long videos. In: Proceedings of the IEEE/CVF International Conference on Computer Vision, pp. 13767–13777 (2023)
10. Xiao, J., Yao, A., Li, Y., Chua, T.S.: Can I trust your answer? Visually grounded video question answering. In: Proceedings of the IEEE/CVF Conference on Computer Vision and Pattern Recognition, pp. 13204–13214 (2024)

11. Zheng, M., Gong, S., Jin, H., Peng, Y., Liu, Y.: Generating structured pseudo labels for noise-resistant zero-shot video sentence localization. In: Proceedings of the 61st Annual Meeting of the Association for Computational Linguistics (Volume 1: Long Papers), pp. 14197–14209 (2023)

The manufacturer's authorised representative in the EU is Springer Nature Customer Service Centre GmbH, Europaplatz 3, 69115 Heidelberg, Germany. If you have any concerns regarding our products, please contact ProductSafety@springernature.com

Printed and bound by CPI Group (UK) Ltd, Croydon, CR0 4YY

30/09/2026

02225577-0012